Practical Country Living

Choosing a Breed of Cattle

5 Needs and 40 Breeds for Selecting Cattle That Fit Your Purpose

Practical Country Living

Choosing a Breed of Cattle

5 Needs and 40 Breeds for Selecting Cattle That Fit Your Purpose

Michelle Lindsey

Homestead on the Range
Abundant Living in Flyover Country

Copyright © 2019 by Michelle Lindsey

No part of this publication may be reproduced, stored in a retrieval system, distributed, or transmitted in any form or by any means, including, but not limited to, photocopying, recording, scanning, or other electronic or mechanical methods, without the prior written permission from the publisher, except in the case of brief quotations embodied in critical reviews and certain other noncommercial uses permitted by copyright law.

Published by Homestead on the Range
Derby, Kansas
https://homesteadontherange.com/

Homestead on the Range is a Kansas-based small business dedicated to serving country living enthusiasts by supplying them with the innovative resources that they need to succeed. Whether your family's farm or ranch is 5 acres or 500, a business or a hobby, in Kansas or in some other part of the world, our goal is to keep you informed and inspired.

Disclaimer
This publication contains materials designed to assist readers and for educational purposes. While the author and publisher have made every attempt to verify that the information provided in this book is correct and up to date, neither the author nor the publisher assumes any responsibility for any error, inaccuracy, or omission.

ISBN 978-0-9975261-3-4 (paperback)
ISBN 978-0-9975261-4-1 (eBook)

TABLE OF CONTENTS

	Introduction	1
1	Choosing a Breed Based on the Five Needs	5
2	Are Purebred Cattle Right For You?	19
3	Types of Cattle Breeds	31
4	Beef Breeds	51
5	Dairy Breeds	85
6	Dual-Purpose Breeds	103
7	Honorable Mention	125
	Appendix: Helpful Resources	131

INTRODUCTION

Part of the charm of a larger homestead is that there is room for cattle. But what kind?

The answer to this question will depend primarily on why you want to keep cattle. Do you want fresh milk? Grassfed beef? Pulling power in the field? Some of each?

When choosing a breed of cattle, there are other factors to consider, as well. Each breed has its own advantages and disadvantages, and what will work well for one family probably won't suit another.

This Practical Country Living guide is designed to walk you through the process of choosing a cattle breed. By the time you have finished the book, you should have:

- Clear expectations for your cattle.
- An idea of whether purebred or crossbred cattle will best meet your needs.
- A handle on characteristics to look for to find cattle that will thrive in your environment and production system.
- A list of a few breeds that have potential.
- Resources to help you make your final choice and purchase your first cattle.

The first chapter introduces the five needs every prospective cattle owner has. The five needs are purpose, scale, environment, marketing plan, and personal preference. The best cattle breed for you will be the one that most closely matches your needs. This chapter will help you take inventory of those needs.

The second chapter discusses the process of creating a breed, an important concept to understand before making your choice. You will learn how both natural and artificial selection shape the genetics of cattle breeds to create populations suited for specific circumstances. You will even have an opportunity to compare and contrast purebred and crossbred cattle with a view toward determining which is best for you. But even if you decide on crossbred cattle, breed composition is still something to consider, and this chapter will tell you why.

The third chapter compares the different types of cattle breeds. You will learn how different genetic backgrounds suit breeds to different environments and production systems. You will also learn about the five major geographic groups of cattle seen in America today, and you will compare the merits of modern and heritage breeds.

Once you have a clear picture of your needs and expectations, it is time to start researching the breeds themselves. The next four chapters offer a total of 40 profiles of beef, dairy, and dual-purpose cattle breeds, covering each breed's origin, genetic structure, population, uses, weight, preferred climate, temperament, health, and production characteristics. These profiles are by no means intended to provide a complete picture of each breed's pros and cons—you will definitely want to consult more comprehensive resources before you make your final decision. However, by reading the profiles, you should be able to quickly zero in on a few breeds to research further.

The final section of the book is devoted to helpful resources. Once you have whittled down your options, read the books and websites listed in this section to find out more about the potential candidates. These resources include more in-depth information, and many provide links to relevant breed associations. Additional resources are also included to provide you with information on caring for your first cattle.

As you can see, this Practical Country Living book is not intended to supply you with all the information you need to keep cattle. Instead, it is intended to assist you with the process of setting clear expectations for your cattle and finding a breed that will meet those expectations. This is an essential process to complete before purchasing animals. The time and care you put into this process should amply repay itself in a herd that fits your circumstances and that you enjoy working with.

CHAPTER 1
CHOOSING A BREED BASED ON THE FIVE NEEDS

There are literally thousands of cattle breeds worldwide. While most Americans won't have to consider this incredible array of choices, they do have to choose from at least several dozen options. So how do you settle on one breed?

The first step, before even looking at the breeds out there, is to know roughly what it is you are trying to accomplish:

- Are you interested in beef, dairy, or multipurpose cattle?
- On what scale?
- In what environment?
- To be marketed in what way?
- To fit what personal expectations?

This brings us to the five needs that your cattle will need to meet:

- Your purpose.
- Your scale.
- Your environment.

- Your marketing plan.
- Your personal preferences.

The better your cattle match and fulfill these five needs, the happier you will be with your choice of a breed.

As you work through this chapter, start a list of the goals and expectations you have for your cattle. Once you have finished the chapter and know roughly what you are aiming for, go back over your list and prioritize your requirements. Armed with this information, you will be ready to research the many breeds of cattle that exist.

Your Purpose

Why do you want to keep cattle? Does your vision involve fresh milk? Grassfed beef? Draft oxen? A combination of these things?

Purpose is probably the most important consideration when choosing a breed of cattle, as it will narrow the field of candidates dramatically.

The three main purposes for which cattle are kept are:

- Beef.
- Dairy.
- Draft work.

Some cattle are best suited for only one of these purposes, making them ideal for specialized operations. Other breeds fit into two or even all three of these categories.

If you will be marketing your cattle through conventional channels, you will want to choose a specialized beef or dairy breed, or perhaps a carefully selected cross of specialized breeds. Agripreneurs seeking to produce grassfed beef or milk will also likely prefer a traditional breed that has been specialized for one purpose.

A homesteader simply looking for a little homegrown food for family purposes may want to consider a dual-purpose breed. A good dual-purpose breed will produce food just as high (or higher) in quality as its specialized counterparts, but at the expense of lower production levels. This can actually be beneficial in a homestead setting where commercial production levels are often overwhelming. As an added bonus, many dual-purpose breeds make good draft oxen, too.

Agripreneurs who plan to sell both beef and milk will need to evaluate whether lower production levels will be acceptable before settling on a dual-purpose breed. Even so, there are breeds that might just fit the bill.

While beef, dairy, and draft power are the primary products that cattle produce, there are other possibilities, as well. Shaggy breeds can produce valuable hides. Highland cattle produce fiber that can be spun into yarn. Horned breeds make fine trophies or polished skulls at the end of their lives. Athletic breeds of Spanish origin are sought by rodeo contractors. And many breeds with strong foraging instincts can be used for land clearing and management with a little know-how. If any of these unconventional uses for cattle are important to you, be sure to add them to your list of requirements.

Questions to Ask

- What is the primary product you want your cattle to produce?
- What additional products would you like your cattle to produce?
- Is a specialized beef or dairy breed most likely to meet your requirements, or would a dual-purpose breed be a better fit?

Your Scale

Closely related to purpose is scale. How many acres will you have access to? What is the stocking rate in your area? How many cattle will you need to keep to turn a profit or to meet your family's milk and meat requirements? These are questions that must be answered when choosing a breed. The more commercialized your operation, the more space it will likely require.

The Question of Size

Many of these questions of scale are also related to the question of cattle size. Not surprisingly, there are many different schools of thought on the subject of size in cattle. The reason for the conflicting opinions is that a compromise is involved in this choice. Large-framed cattle weigh more at slaughter; on the other hand, they take longer to grow and may never reach their full potential without large amounts of feed. Small-framed cattle are unquestionably more efficient, requiring fewer inputs and maturing quickly. However, they can put on undesirable amounts of fat if overfed, will produce fewer total pounds of beef, and can be harder to market through conventional channels, depending on current market demands.

Many people in harsh climates prefer smaller cattle because they can maintain healthy weights—even when the going gets tough—simply because they don't need to eat as much as their larger counterparts. More hospitable conditions cause more controversy. On the one hand, producers frequently don't see the value of raising small cattle in a favorable climate when the sale barn tends to reward size. Even when direct marketing grassfed beef, larger cattle can be considered valuable for their heavier weights and leaner beef. On the other hand, some cattle raisers still favor small to medium-sized cattle in wet, mild climates because they make higher stocking rates possible, increasing pounds of beef production per acre without supplemental feed.

One tool that can make it easier to whittle down your options if you are considering beef cattle is the frame score. This scoring system was developed to evaluate the skeletal size of an animal based on hip height and adjusted for age. In most cases, a frame score will not change much as a bull, steer, or heifer grows, making it useful for roughly estimating the mature size of an animal.

Cattle are scored by measuring their height at the hip. A table (typically available from the appropriate breed association) is then used to interpret the measurement based on the animal's age.

Frame scores can theoretically range anywhere from 1 to 11, 1 being extremely short and 11 being extremely large. In practice, however, extreme frames are rarely seen except among miniature breeds (more on that in a moment). Most standard-sized cattle rank anywhere from 2 to 9. British breeds, such as the Angus, tend toward the smaller end of the spectrum, while breeds from Continental Europe, such as the Charolais, tend toward the larger end. (Keep in mind that there is a great deal of variation within each breed.)

Probably the most important factor in deciding on an ideal frame score for your cattle is your marketing plan. Commercial cattle marketed through the sale barn do not have to be extremely large. Cows with frame scores of 6 and above tend to be prone to reproductive issues, and the commercial market will usually accept medium-sized calves. A frame score between 5 and 6 is typically favored for the commodity market. A cow in this height range will typically weigh somewhere between 1,175 lbs. and 1,250 lbs. when mature, while a steer this size will be ready for slaughter after reaching 1,150 lbs. to 1,250 lbs.

Grassfed cattle are another subject. Smaller cows and steers have proven themselves again and again in this niche because they fatten well on grass alone and reproduce consistently with few inputs. Frame scores between 2 and 4 are common, representing a weight range of 955 lbs. to 1,100 lbs. for mature cows and 850 lbs. to 1,050 lbs. for steers ready for slaughter.

Not all producers agree with these two solutions, of course. Some commercial cattle raisers feel that they still have a more favorable profit margin by reducing costs with small-framed cattle. On the other hand, some grassfed producers in mild climates see no reason why they shouldn't raise larger cattle with heavier carcasses.

The frame score is simply a tool. There is no one ideal score to aim for. Each cattle owner must set his own goals and choose a range of frame sizes that will help him reach those goals.

The odds are pretty good, however, that most future cattle owners should be considering a small to medium animal, no matter what they intend to produce or how they intend to sell it. Smaller cattle are far more efficient than larger cattle, especially when raised on forage. Not only are they more capable of feeding themselves with energy to spare for reproduction and lactation on natural resources alone, but nearly every pasture can support a greater number of small cattle than large cattle.

This does not necessarily mean that you have to consider miniature cattle—it just means that in most cases you will want to avoid truly large breeds, and you will probably want to seek out more moderate-sized cattle within the breed of your choice.

Pros and Cons of Miniature Cattle

Miniature cattle are a good option for some situations. Miniature cattle generally have a frame score of 1 or less, with a possible range of 1 (on the bigger side of little) down to 000000 (truly tiny). A mature bull with a frame score of 000000 measures only 36 inches at the hip!

Here are a few of the advantages of keeping miniature cattle:

- **Small-acreage solution.** Yes, miniature livestock fit well into small spaces! This is probably the most important reason to consider a mini breed. Of course, you will still need to check your local zoning regulations to make sure livestock of any kind are legal in your area.
- **Low feed requirements.** Miniature cattle need less hay and pasture than their larger counterparts.
- **Less manure.** Especially important in small areas.
- **Easy handling.** Most miniature animals are a little easier for children to work with. Many adults appreciate this advantage, too.

- **Homestead-sized production.** It can be difficult finding room in the refrigerator or freezer to store milk and meat. Most miniature cattle breeds produce just the right amount of food for homesteaders. A miniature beef steer will even fit comfortably into a standard upright freezer.
- **Just plain unique!** What's wrong with choosing a miniature breed just because it's different? Miniature animals can make an interesting addition to your family farm.

But there are pitfalls to be aware of, as well:

- **The fad factor.** Miniature livestock breeds are currently a fad, attracting some unscrupulous breeders to the market. Of course, responsible breeders do exist, and they offer animals with great potential value to homesteaders. But the old adage holds true here: "Let the buyer beware!" Make sure your miniature cattle breeder of choice consistently selects for good health and temperament.
- **Price.** Because mini breeds are in short supply and high demand, the prices asked for quality breeding animals can be staggering. A good rule of thumb is the lower the frame score, the higher the price tag.
- **Genetic defects.** There are some defects associated with small size in animals, most notably dwarfism. Dwarfism may result in a malformed skeleton or difficulties breeding and giving birth. It is wise to buy miniature livestock from a source that checks for such defects.
- **Vulnerability.** Miniature livestock can be easy pickings for predators. Their small size makes it harder for them to defend themselves. Also, any mini breeds kept with larger animals can be seriously injured by their pasture companions when playing or fighting.
- **Milking logistics.** Milking a miniature cow can be a problem for a taller person, as it may involve a choice between sitting on the floor and training the cow to stand on an elevated

platform. Also, a few miniature cows have very small teats that are hard to milk by hand.
- **Breeding challenges.** Miniature livestock should never be crossed with standard-sized breeds, which can make it harder (and more expensive) to get your tiny cows bred. In addition, smaller size may make artificial insemination more difficult with mini breeds.

One of the keys to avoiding many of the difficulties listed above is to purchase miniature livestock from a reliable source. Do your homework! Stay away from fad breeders out to make easy money. Go with someone who is dedicated to providing a valuable service to country families. This will help ensure that your miniature cattle will provide you with years of enjoyment.

Milk Production

For those who are considering milk rather than beef, looking at scale will involve estimating desired milk production. Obviously, the homesteader just wanting farm-fresh milk will have very different requirements than the aspiring entrepreneurial cheesemaker.

Beginning dairy farmers may be surprised to learn that milk is typically measured in pounds rather than gallons. This is because a gallon is a measure of volume and can be influenced by the frothiness of the milk. The pound is the preferred measurement due to its accuracy—a pound of milk is a pound of milk no matter how foamy it is. As an added plus, dealing with pounds is far easier than trying to measure and add up fractions of gallons.

One piece of equipment that you may want to keep in your dairy parlor is a scale. But for everyday home use, you can simply convert back and forth by remembering that a gallon of fresh milk weighs 8.6 pounds on average. (Some homesteaders simplify even further by rounding down and reciting, "A pint's a pound the whole world 'round.")

So how many pounds of milk will you need to produce in a year? If your milk will be for home use, you should be able to calculate your dairy needs by noting how often you buy a gallon at the store. You can also make an estimate based on the fact that the average American consumes 20 to 25 gallons of milk each year. This translates to around 170 to 215 pounds of milk annually.

Now if you plan to make dairy products in addition to drinking fresh milk, your milk requirements will increase considerably. Consider some of these facts:

- It takes a pound of milk to make a pound of yogurt.
- It takes about 10 pounds of milk to make a pound of cheese.
- It takes about 12 pounds of milk to make a gallon of ice cream.
- It takes about 22 pounds of milk to make a pound of butter.

Many homesteading families like to have enough milk for making cheese and butter weekly, drinking regularly, and giving away to friends and family on occasion. For this scale, 2 to 3 gallons daily is usually the ideal level of production. This translates to about 5,000 to 8,000 pounds of milk annually. Note that the average commercial dairy cow produces about 21,000 pounds per year! Obviously, the homesteading family will want to consider a breed with less milking potential, perhaps even a dual-purpose breed.

If you are considering starting a dairy farm for profit, things get more complicated. How much milk you will need to produce will depend on your customer base, and that is a project for serious market research. Once you have estimated the demand, however, you should be able to use the statistics above to calculate how many pounds of milk you will need to produce annually to supply your customers.

Questions to Ask

- How many acres do you have access to?
- What is the recommended stocking rate in your area?

- How many cattle do you plan to start out with?
- How much do you expect your herd to expand over the next few years?
- Will you be producing milk or meat for sale or for home use?
- What frame score and weight range will best meet your requirements?
- How many pounds of milk do you need to produce annually?

Your Environment

Your environment is also a crucial consideration, as it cannot be changed except at great expense. The better your cattle are adapted to your environment, the fewer inputs it will take to keep your herd productive and healthy.

Take stock of your environment:

- **Climate.** Is your weather generally cold or hot? Humid or arid? Some combination of extremes?
- **Altitude.** Do you live at a high altitude? (High altitudes are hard on the lungs, arteries, and hearts of cattle from unadapted genetics.)
- **Forage.** How good is your forage? Are your native grasses lush or sparse? Do you have access to tame pasture plants? How does the nutritional value of your forages vary seasonally? Are there known challenges associated with the species of forages you have access to, such as bloat or fescue toxicosis?
- **Parasites.** Do you have a lot of parasites in your area? If so, what kinds?
- **Disease.** Is there any disease pressure in your area? Infectious diseases? Climate-related diseases, such as cancer eye or foot rot? Diseases related to mineral deficiencies?

One thing you will notice as you research cattle breeds is that heritage breeds tend to adapt better to harsher conditions. Many modernized breeds have been selected to produce large quantities of milk or meat in exchange for a tightly controlled environment, including ample feed and medical attention. If you cannot or choose not to replicate these commercial conditions, you will probably be much happier with a breed that knows how to rustle its own living in a natural environment.

Of course, no one breed, heritage or otherwise, can adapt to all conditions. Some breeds are better suited to hot, dry climates, while others were made to tolerate cold, wet climates. It is important to select a breed that will thrive in your area.

Questions to Ask

- What type of climate do you have in terms of both temperature and precipitation?
- Do you live at a high altitude?
- What types of native forages do you have access to?
- What types of tame pastures do you have access to?
- How does the nutritional quality of your forage resources vary seasonally?
- What challenges, toxicity risks, and nutritional deficiencies are associated with your forage resources?
- What types of parasites are common in your area?
- What diseases are common in your area?
- Are there any other environmental idiosyncrasies to take into consideration in your area, such as intense sunlight, rough terrain, or muddy pastures?

Your Marketing Plan

What product or products will you be marketing? You have already considered the differences between beef, dairy, and dual-purpose breeds. However, there are quite a few options within each of these categories.

Those who keep beef breeds can choose from a wide variety of emphases, programs, and options. The following is just a sample:

- Organic beef.
- Cattle hides.
- Roping steers.
- Heritage beef.
- Purebred seedstock.
- Certified Angus beef.
- Gourmet grass-finished beef.
- Commercial crossbred steers.
- Calves for children's projects such as 4-H.

Dairy producers have other options:

- Butter.
- Yogurt.
- Ice cream.
- Show stock.
- Draft steers.
- Organic milk.
- Grassfed milk.
- Artisan cheese.
- Crossbred calves for beef.
- Easy-to-digest A2 beta casein milk.

Those who raise dual-purpose breeds will be able to select from some combination of these two lists.

Different breeds lend themselves better to different marketing systems. For example, those who plan to sell gourmet beef to fancy restaurants would want to look at breeds noted for tenderness and flavor. Someone with an interest in cheesemaking would compare the protein and butterfat levels of various dairy breeds to select one that will be able to produce the desired results.

Even if your cattle will be strictly for your family's use, you will still want to take a close look at the quality of meat and milk you intend to raise to determine the breeds that will best meet your needs.

As a final note, while you are working out your marketing plan, be sure to set some budgetary limits. Having a clear idea of how much you are willing to spend on purchasing cattle and having them delivered will affect your choice of breeds. Rare cattle, breeds with limited availability in your area, and animals from bloodlines in high demand can be rather expensive.

Questions to Ask

- Will you primarily be producing cattle or cattle-related products, such as beef and milk?
- How will you add value to your cattle or products?
- What marketing channels will you use?
- Who will be your primary customer?
- What are your customers' expectations for quality and production levels?
- What are your expectations for quality and production levels?
- What benchmarks will you use to evaluate quality and production levels?
- What is your startup budget?

Your Personal Preferences

Once you have a list of breeds that will be viable in your scenario, the final choice should factor in your personal preference. After all, you are planning on spending time with your cattle every day for years to come—you had better enjoy that time!

One of the most important things to think about when it comes to personal preference is temperament. Few people enjoy working with nervous or aggressive cattle. But between those two extremes lies a wide range of personalities, varying from smart and active to laid-back and easygoing. Make sure you end up with cattle that won't overwhelm you with their quick thinking or disgust you with their blank stare, depending on your preference.

Make sure you truly love the breed you have chosen. Does its story appeal to you? Will it look attractive in the front pasture? Do you find yourself reading up on that breed to the exclusion of all others? These seemingly trivial considerations are often a good measure of how much you will enjoy raising a given breed.

Keeping cattle should be fun—selecting a breed you love is the first step toward that goal.

Questions to Ask

- How much experience do you have handling cattle?
- How comfortable will you be working in close proximity to large animals?
- Is trainability an important consideration in your situation?
- Will protective maternal instincts be necessary to keep calves safe from predators?
- Will you be keeping a bull?
- Is there a breed or group of breeds that appeals to you because of its history, appearance, or other unique characteristics?
- What traits will you value most in your cattle?

CHAPTER 2
ARE PUREBRED CATTLE RIGHT FOR YOU?

As with most things, pure breeds come with their pros and cons. Whether or not a pure breed is right for you will boil down to the specifics of your five needs. Depending on the circumstances, sometimes a pure breed will fit better and sometimes a crossbred animal will fit better.

What Makes a Breed?

Before choosing between purebred and crossbred cattle, it is best to have a clear picture of what a breed is and what makes it pure.

A breed is a group of related animals that strongly resemble each other in genetic makeup and thus in appearance. They also tend to have similar temperaments and production traits. The animals that make up the breed, when mated together, produce offspring that resemble themselves.

The families that make up breeds are collected through selection. Selection is the process of keeping some animals for breeding purposes (thus perpetuating their genetics) and rejecting others (thus eliminating their genetics from the gene pool). Selection can be made naturally or artificially.

Selection perpetuates some genes at the expense of others. As this process occurs over several generations of cattle, the population starts to become more uniform in the traits for which they have been selected, whether that is appearance or simply the ability to survive in a particular environment. This uniformity is the hallmark of a pure breed.

Of course, not every breed is uniform in every trait. Some traits in some populations receive very little selection pressure. To take a hypothetical example, if a group of cattle is thrown together without regard for color and is selected strictly for beefy build and polling (hornlessness) over many generations, the resulting breed may not be uniform in color at all. But it should be so uniform with regard to the traits of beefy build and polling that it can readily be identified as a distinct population nevertheless.

Which traits receive the most selection pressure depends on whether selection is primarily natural or primarily artificial.

Natural Selection

Natural selection is carried out by the environment. Examples of natural selection would include a drought that starves all but the most forage-efficient animals, a disease that kills off all but the strongest, or a predator attack that eliminates the offspring of all but the most vigilant mothers.

Just as selection can be made by natural forces, some cattle breeds (particularly feral breeds) are products of the environment, having been shaped for centuries by the prevailing conditions of climate, terrain, forages, and more. A naturally occurring breed remains pure primarily due to geographical isolation.

Artificial Selection

Artificial selection is carried out by humans. Artificial selection includes both the process of choosing and purchasing new breeding stock and the process of removing undesirable breeding stock through sale, slaughter, or castration.

This process of creating breeds has been carried out ever since cattle have been domesticated. The earliest populations of cattle that could be considered breeds were typically selected for factors of importance in religious and cultural contexts. For example, African tribesmen have long bred cattle to be premier status symbols by selecting for large horns or flashy colors. Breeding cattle for production traits—whether that is beefy build, copious milk production, strength for draft work, or some combination thereof—occurred later as peoples came to depend on cattle for food and power instead of social standing. As time went on, cattle breeds were increasingly bred for high levels of production in very specific applications.

There is a common notion that the process of creating breeds dates back to the 1700s and the intense inbreeding practiced by Robert Bakewell, thus making breeds a recent invention. While Bakewell certainly revolutionized the process by which modern breeds have been shaped, going back to the basic definition of a breed shows us that cattle breeds have existed ever since man has raised domestic cattle.

Most breeds today are developed through artificial selection, being the result of an effort to propagate livestock that possess a given set of desirable characteristics, such as appearance or production. Artificially selected breeds are largely kept pure by man-made limitations. For instance, producers might agree that their genetics are satisfactory and formally designate their livestock as a pure breed. Or, in some countries, cultural barriers may restrict peoples from exchanging cattle, thus resulting in distinct breeds.

More than a few breeds are created by the combined efforts of man and nature. Good cattle breeders frequently mimic natural selection processes by culling cattle that fail to thrive in their environment without expensive inputs.

Advantages of Purebred Cattle

The main advantages of purebred cattle are in keeping with the reasons breeds evolve in the first place. They include consistency, the ability to take advantage of heritability, and leverage for improving a herd.

Consistency

Consistency is a major benefit to the new cattle owner, as it reduces the risk of surprises (poor disposition, genetic health problems, etc.) during the critical learning stage.

But consistency is also important any time a cow herd is being maintained. For example, if you are planning on starting a commercial cow-calf herd, you will appreciate raising calves that are consistently low-maintenance while still meeting sale barn demands. If you intend to market grassfed beef, you will probably enjoy the best success by raising your own steers from genetics already selected for consistent tenderness and easy fleshing. If you are interested in keeping a dairy cow, when it comes time to breed her, you may want to consider consistency for the sake of producing a useful calf.

Purebreds have an advantage over crossbred cattle in consistency. This comes from having many like genes paired together so that the offspring in turn receive many pairs of like genes.

This is not to say that crossbred animals can never be consistent. The Angus and the Hereford are crossed to produce the Black Baldy steer for commercial beef production time and time again for this very reason. But the consistency is due to the fact that the parent breeds are both uniform and consistent. They also have just the right combination of genes to produce a fast-growing beef animal when mated together. But when a Black Baldy is used to produce a second generation, this consistency is immediately lost.

For this reason, if you intend to raise your own calves, you will likely want to take advantage of the consistency of either a pure breed or a proven combination.

Heritability

Heritability is an estimate of how much a given trait is determined and influenced by genetics versus the environment. A highly heritable trait is one that is primarily controlled by genetics, such as coat color. A trait that is not highly heritable is one that is more heavily influenced by environmental conditions like nutrition and sanitation, such as immune response to pathogen pressure.

Even if you have no intention of raising your own calves, you may find that certain highly heritable traits are important to you. Examples of important traits with high heritability in cattle include:

- Disposition.
- Resistance to horn flies.
- Birth weight.
- Pre-weaning weight gain.
- Hip height at weaning (useful for calculating frame score).
- Post-weaning weight gain.
- Yearling weight.
- Mature weight.
- Male libido.
- Pelvic area (related to calving ease).
- Length of teats (excessively small teats have poor milk flow, while excessively large teats are hard to suckle and prone to infection).
- Retail yield of carcass.
- Carcass loin eye and ribeye area.
- USDA carcass grade.
- Meat tenderness and juiciness.
- Milk protein percentage.
- Milk butterfat percentage.

Purebreds have an advantage when it comes to highly heritable traits, as these are the traits that respond quickly to selective breeding. Therefore, if a highly heritable trait is on your priority list, a pure breed already selected for the desired trait will probably fit your requirements well.

If you will be raising your own calves, heritability becomes even more important. For example, if you intend to raise steers on pasture and direct market the beef, meat tenderness is likely to be of vital importance to you. Therefore, you may want to consider a pure breed with a good track record for tender beef. Not only will you start out at an advantage, if you do feel the need to improve the tenderness of your beef calves, selective breeding should bring you to your desired result fairly quickly.

Herd Improvement

Choosing a pure breed gives you a head start on the herd improvement process should you choose to raise calves. There are enough cattle breeds in the United States today that it is easy to find one that will fit your environment and purpose, reducing the amount of time that it will take you to select for animals that meet your objectives.

However, improving a herd does require you to be choosy with the breeding stock that you purchase. Truly good purebred breeding stock is not cheap. In fact, the better the quality of the animal, the heftier the price tag it will generally command. The temptation, therefore, is to compromise and start out with inferior breeding animals. These will in turn beget inferior animals. The reputations of entire breeds have been tarnished this way.

Also, if a breeder is not foresighted during the selection process, unnatural extremes can be bred into the herd. This is one reason that crossbreds sometimes display an advantage over purebreds—the extremes tend to moderate, depending on the breeds that are selected to produce the cross (more on that later in this chapter). However, a breeder is hard-pressed to go wrong selecting for cattle that remain healthy and productive in a given environment with minimal inputs, and this can be done with purebred cattle just as easily as (if not easier than) it can with crossbred cattle.

In fact, it is even possible for those with the time, resources, and determination to take advantage of the higher inbreeding levels of purebred livestock to improve the breed. Remember, pure breeds have large numbers of like genes paired together. Also bear in mind that every animal in the world has a high likelihood of having at least one genetic defect somewhere in its makeup. When such deleterious genes are dominant, or producing effects that the other gene in the pair cannot mask, they are quickly identified and removed from the breeding population. Therefore, most of the deleterious genes that give cattle breeders trouble are recessive, or masked by the effects of the other gene in the pair. Inbreeding depression occurs when such deleterious genes pair up and produce negative effects in the animal, typically seen in the areas of health and reproduction.

The practice of crossbreeding seeks to always mask these deleterious genes under beneficial ones, producing animals that are economically viable. With purebred livestock, permitting the recessive genes to pair up unmasks them, providing an opportunity to identify and cull the carriers. If you don't want to worry about the effects of genetic defects in your cattle, crossbred animals may be right for you. If you are interested in searching out and eliminating the defects altogether, working with purebred animals will help you achieve your goals with greater success.

Unlike crossbreeding, pure breeding is not a quick fix. Achieving the desired results requires proper selection of breeding stock over time, as well as stringent culling of undesirable animals. That said, if you have the time and patience to develop great genetics, whether for your commercial cow-calf business or for the purpose of achieving superior grassfed beef or dairy products, purebred cattle are the way to go.

Advantages of Crossbred Cattle

But there are also many instances where crossbred cattle have a decided advantage over purebred cattle. The primary advantage of crossbreeding is hybrid vigor. However, the ability to create a custom blend of genetics can also be advantageous in some situations.

Hybrid Vigor

Hybrid vigor occurs when many unlike genes are paired together. Practically speaking, this translates into a number of characteristics desirable in many situations:

- Enhanced health, particularly at birth.
- Improved reproductive performance.
- Increased production.

The traits that respond to a boost in hybrid vigor are typically those that are not highly heritable. These traits include:

- Longevity.
- Foot and leg structure (important for avoiding lameness).
- Disease resistance.
- Incidence of metabolic disorders.
- Age at puberty.
- Fertility.
- Calving interval.
- Incidence of retained placenta.
- Incidence of uterine infection.
- Intensity of beef flavor.
- Udder size and shape (important for udder health).
- Lifetime milk yield.
- Milk somatic cell score. (Somatic cells are body cells, primarily white blood cells; when the cell count in the milk goes up, it typically indicates stress or poor health.)

Because hybrid vigor occurs when unlike genes are paired together, it follows that hybrid vigor is maximized when as many unlike genes are paired together as possible. When crossbred cattle are interbred, the genes recombine and frequently form like pairs again, thus reducing the hybrid vigor levels. This is why crossbreeding is typically not recommended after the first generation.

If you have no intention of raising calves, but are interested strictly in production, crossbred cattle can be a good fit due to their hybrid vigor. If you do intend to raise calves, such as steers to market at the sale barn, you can still make the most of hybrid vigor by keeping purebred cows, using a purebred bull of a different breed, and buying rather than breeding replacement heifers as needed.

Customized Genetics

Crossbred cattle can excel where a precise combination of highly heritable and less heritable traits is required to meet market specifications. Take, for instance, the Black Baldy. The parent breeds (the Angus and the Hereford) have been selected for weight and USDA grade, both heritable traits that will be transmitted to their offspring. Crossing the two breeds to cash in on hybrid vigor improves the growth rate and often the health of the calves—definitely an advantage in the feedlot.

Another situation where crossbred cattle might have a useful advantage is when a breed that meets your needs is not readily available. There are well over a thousand cattle breeds in the world, but this will not help you if you cannot obtain quality animals at the right price. When this is the case, crossbreeding may be a better choice than compromising by selecting a breed that is not particularly suited to your requirements.

But this does not make crossbreeding a silver bullet—a way to create cattle that have it all. There is no such thing as a silver bullet. There are always trade-offs to be made when purchasing or breeding cattle: Do you want a steer that will fatten quickly or one that will excel in beef quality? Do you want a cow that will produce large quantities of milk or one that will offer high levels of protein, vitamins, and minerals in her milk? Do you want an animal that can excel in one area of production or one that can be a multipurpose homestead animal? Do you want a cow that is high-producing or highly forage efficient? As long as you are realistic in your expectations, however, the ability to combine breeds can be useful in achieving just the right genetic blend.

Given the amount of knowledge and care required to develop such a genetic blend, this is not a project most beginners are willing to tackle. Instead, you will probably do well to learn which crosses are prevalent in your area and work with a proven combination.

Why Breed Choice Matters in Crossbreeding

But even if you do decide to go with crossbred cattle, you will still want to consider the breeds that will go into the mix.

Remember, hybrid vigor is maximized when as many unlike genes are paired as possible. Translated into practice, this means that hybrid vigor is maximized when the two breeds being crossed are quite uniform, but quite different from each other. Crossing two British beef breeds, such as an Angus and a Hereford, produces a calf with plenty of hybrid vigor. Crossing an Angus with a Brahman—an Indian breed with a very distinct genetic background—provides even more hybrid vigor.

Another thing to consider is complementarity. There are specific crosses that tend to work consistently and well in the world of cattle. This is often because one breed will bring something to the table that the other breed lacks. If we find one breed that promises to meet our needs well in carcass quality but that is lacking in growth rate, we can compensate for this shortcoming by ensuring that the other breed we choose is known for good growth rates. This is the reason that the Angus, known for marbling, is frequently crossbred with big, fast-growing but very lean beef breeds from Continental Europe to produce commercial beef calves.

Of course, we must be careful to avoid introducing unwanted problems, as well. This is why, even though the Angus x Brahman cross provides more hybrid vigor than the Angus x Hereford cross, the Brahman cross is not too common outside of the Deep South, where its tolerance of heat, humidity, and parasites is absolutely essential. The Brahman component tends to introduce the likelihood of tough beef to the equation, as well as the potential for unsound temperaments.

Thus crossbreeding and hybrid vigor do not automatically eliminate problems from the herd. If two cattle breeds that do not suit the environment or production system are crossbred, the result will be more cattle that do not suit the environment or production system.

CHAPTER 3
TYPES OF CATTLE BREEDS

Whether you intend to keep purebred or crossbred cattle, the next step in the process of choosing a breed or breed combination is to start narrowing your options. A breed that will meet your five needs will likely have been developed in a time, place, and manner consistent with your objectives. Therefore, before looking at the individual breeds, we will examine the ways in which those breeds came about.

Genetic Background

Breeds that have been created through different mechanisms vary in important ways. For instance, a feral breed that has been shaped solely by the environment for centuries is doubtless extremely hardy; on the other hand, it likely has a difficult temperament that makes it less than useful to mankind. In contrast, a breed developed primarily by humans will probably be selected for good disposition, even at the expense of hardiness.

Livestock breeds can be divided into five groups based on their genetic origin and the focus of selection:

- Industrial.
- Standardized.
- Composite.
- Landrace.
- Feral.

Industrial Stocks

Industrial stocks are highly specialized, tightly controlled breeds typically created and maintained by corporations. While there are exceptions, some of these breeds may never enter the hands of private owners. They are created for one purpose—to provide food and plenty of it.

The gene pool of an industrial stock is frequently chosen from a small, high-production segment of a standardized breed. Subsequent breed development involves intense selection pressure focused on maximizing production traits. As a few high-production sires emerge via progeny testing, artificial insemination spreads the genetics of this handful of animals across the globe. From a small genetic foundation, a massive and remarkably uniform animal population is thus created to supply consumers worldwide.

Industrial breeds are typically high-octane animals that are expected to produce massive amounts of milk or meat in exchange for scientifically formulated rations. They usually spend their entire existence in close quarters, as confinement allows for tighter control of the production process.

This model of raising industrial stocks is standard in the poultry and swine industries in the United States, but is not typical in the beef cattle industry. The dairy industry, however, possesses one industrial breed—the Holstein, selected for amazing quantities of milk.

To summarize:

- Industrial stocks are developed exclusively through artificial selection.
- A small subsection of a standardized breed typically forms the genetic base.
- The selection process is guided primarily by corporations.
- Selection is focused solely on production levels.
- The gene pool is extremely narrow.
- The populations are very large.
- Geographic distribution ranges from limited and exclusive to global.
- The primary advantage is very high production levels.
- The primary drawback is very high maintenance requirements.

Standardized Breeds

A standardized breed, as its name suggests, is bred to meet certain standards, whether those are of conformation, production, or behavior.

The animals that form the foundation of a standardized breed can come from a variety of sources. Historically, most were probably originally landraces for which a standard and a breed association were subsequently formed. Other breeds were originally composites of other breeds that have stabilized and cohered over time.

Standardized breeds are characterized by the organization and coordination of breeders. The breeders typically select for the same standard and record their animals through the use of pedigrees and herdbooks, thus making it easier for all the breeders to find the best seedstock for their programs. The standards and records are maintained by one or more breed associations, which typically also play a significant role in breed promotion.

While the coordinated effort behind a standardized breed tends to result in a uniform population, different breed organizations have different philosophies that allow for some degree of variation through the introduction of outside blood. Some registries are closed, meaning that only the offspring of parents documented back to a certain set of ancestors can be registered and thus allowed to contribute genetics to subsequent generations of the breed. A few organizations may accept unregistered animals that can demonstrate their lineage through visual appraisal, blood typing, or DNA testing. Others allow upgrading. Upgrading is the process of crossing one generation of the desired breed to another breed to broaden the gene pool or introduce new traits, then breeding the crossbred offspring back to the parent breed. After a few generations, there is little difference between the upgraded animals and the original cattle except for lowered inbreeding levels.

Different breeds are backed by different philosophies as to the purpose of the breed—an important consideration when establishing promotion policies. Some breeds are selected for one sole purpose, such as beef production or milk production, while others are bred to be versatile. When the members of a breed association cannot agree on the purpose of the breed, they often divide and form two associations. If this split results in the genetic isolation of the different segments of the breed over the course of several generations, the separate populations will eventually develop into entirely separate breeds.

As one would expect, standardized breeds can be selected in dramatically different directions to accommodate varying purposes. In fact, any given standardized breed has probably been selected for different traits during different eras of its history, depending on what makes the breed marketable at the time. Breeds known for certain conformation characteristics, whether that is the solid black coat of the Angus or the impressive horns of the Texas Longhorn, are bred for the enhancement and perpetuation of those characteristics. Breeds with known temperament issues have been reclaimed through a focus on disposition. Standardized dairy breeds are frequently selected for increased milk production in addition to dairy conformation to make them competitive in the modern economy.

Thus, standardized breeds can be either tightly inbred or highly diverse, extremely specialized or remarkably versatile. The genetic makeup of a standardized breed always has some level of uniformity, or it would not be a breed. However, the level of uniformity depends on the philosophy of the breed association combined with the practices of the individual breeders.

Examples of standardized cattle breeds abound. They range from the Angus (a specialized beef breed) to the Jersey (a specialized dairy breed) to the Dexter (a dual-purpose homestead breed).

To summarize:

- Standardized breeds are developed primarily through artificial selection.
- A landrace or composite breed forms the genetic base of most standardized breeds.
- The selection process is guided partly by the breed association and partly by the individual breeders.
- Selection is focused on a written standard, which may include benchmarks for conformation, temperament, and/or production traits.
- The gene pool ranges from extremely narrow to extremely broad.
- The populations vary from extremely large to extremely small.
- Geographic distribution is typically broad, ranging from regional to global.
- The primary advantage is consistent genetics.
- The primary drawback is the risk of shortsighted or controversial breed fads.

Composites

A composite breed is actually a blend of breeds fixed at a precise ratio. In the development phase, creating a composite breed involves crossing two or more pure breeds repeatedly to achieve this ratio. That may sound simple, but to create a sizable gene pool free from inbreeding problems requires dozens of sires and potentially hundreds of females. The crossbred offspring of this initial foundation have to be crossed and preferably recrossed until the population stabilizes with each breed contributing to the overall genetic makeup in the desired proportion.

Once the composite breed is established, however, management becomes extremely simple. A composite is mated like a pure breed. In other words, composite animals can be freely bred to each other without concern about losing consistency. Also, unlike long-term crossbreeding systems that maximize hybrid vigor through the maintenance of several herds, each of a different breed or cross, only one breeding group is required with composite breeds.

As you might have noticed, a true composite breed is essentially a standardized breed in its infancy. Genetic makeup forms a major part of the standard of most composite breeds, but the written standards for these breeds may also focus on conformation and production traits.

A few populations are considered composites or "open" composites but are maintained with regular influxes of new genetics from other breeds. While this guarantees that inbreeding will be avoided entirely, it adds an element of inconsistency into the population. Open composites tend to fluctuate wildly in performance unless managed carefully and deliberately.

There are several reasons people might opt to use a composite breed or sire:

- Predictability.
- Consistent levels of hybrid vigor—very difficult to maintain in both traditional pure breeding and long-term crossbreeding programs.

- The ability to combine desired traits of several pure breeds, thus quickly creating an animal population ideally suited to a given set of environmental or economic circumstances.
- Ease of management; other types of crossbreeding rotations can become painfully complicated, particularly where the land base limits the number of separate herds that can be maintained.

Those who raise composite livestock must remember that inbreeding is still a potential threat. Mating extremely closely related animals is not recommended, even in a composite population.

For those seeking a balance between consistency and hybrid vigor and those who cannot otherwise find the best genetics for their situation, a well-bred composite may be the answer.

There are many different composite breeds available in the beef cattle world today. Examples include the Beefmaster, the Brangus, and the Santa Gertrudis.

To summarize:

- Composites are developed primarily through artificial selection.
- Multiple standardized breeds typically form the genetic base.
- The selection process may be guided by some combination of research institutions, breed associations, and individual breeders.
- Selection is focused partly on maintaining a specific breed ratio and partly on a written standard, which may include benchmarks for conformation, temperament, and/or production traits.
- The gene pool is very broad.
- The populations are typically moderately large.
- Geographic distribution may be local initially, but typically broadens to at least regional levels.

- The primary advantage is a balance between consistency and hybrid vigor.
- The primary drawback is the difficulty of creating a composite breed that is genetically stable yet free from inbreeding difficulties.

Landraces

A landrace is a group of genetically related animals unique to a given geographical area. Landraces come about over time as animals within the area interbreed for many generations with relatively little outside influence.

The environment is the major factor that shapes the landrace. If a given animal is not suited to local conditions, it probably will not thrive and produce offspring, eliminating its genes from the population. Instead, other animals that can thrive in those conditions will produce the next generation.

But even though nature has the most say about the gene pool of a landrace population, humans also play a role. Indeed, this element of human intervention is part of what separates a landrace from a feral population.

People shape the genetics of a landrace both passively and actively. Passive selection occurs when humans alter the environment or its natural effect on animals through various management practices. These practices can include:

- Providing shelter.
- Providing feed or supplements.
- Restricting animal movement to specific areas.
- Doctoring sick animals.
- Assisting with animal births.

We often consider these practices to be part of good stewardship of our animals, and rightly so. However, the fact remains that most management practices alter the course of nature and enhance the chances of survival and reproduction for animals that otherwise might have left little genetic legacy.

What some animal breeders overlook is the role that active selection has played in shaping many landraces. The histories of many landrace breeds show us that humans have specifically chosen certain animals to reproduce due to their desirable traits and have culled others due to their undesirable traits. Sometimes the entire landrace population is shaped in this way, while in other cases one family of breeders has simply established their own bloodline within the broader population.

Humans have actively selected some landraces for:

- Production.
- Working ability.
- Versatility.
- Specialized purpose.
- Color.

This brings up an important question: If landraces have been actively bred for appearance and production, then is there any difference between a landrace and a standardized breed?

The simple answer is yes.

Note the factors that most landraces have in common:

- Concentration in a specific geographical area.
- Limited numbers.
- Genetic isolation primarily due to geographical isolation.
- No written standard.

Compare and contrast this with the characteristics of a standardized breed:

- Broad geographical distribution.
- Numbers varying from extremely rare to extremely popular.
- Genetic isolation primarily due to attention to pure breeding.
- Written standard of desirable characteristics.

However, now that the Internet has made promoting breeding animals to distant buyers possible, the lines between landrace and standardized breed have become a little more blurred than advocates of either would perhaps prefer.

At present, there is enough difference between a landrace and a standardized breed to say that the landrace is typically the more genetically diverse of the two. There are exceptions:

- A landrace breed teetering on the brink of extinction may lose its genetic diversity through population depletion and subsequent inbreeding.
- A standardized breed may have maintained its original genetic diversity through the preservation of several distinct bloodlines.

Cattle breeds that can still be classified as landraces include:

- Florida Cracker.
- Pineywoods.
- Randall.

However, a landrace does not always have to stay a landrace. If a group of breeders decides to pursue more formal breeding methods, the landrace may become the basis of a standardized breed. The Texas Longhorn is a classic example of a landrace that has been standardized, as are the Corriente and the Lineback.

To summarize:

- Landraces are developed through both natural and artificial selection.
- Multiple breeds from a variety of backgrounds typically form the genetic base.
- The selection process is guided partly by the environment and partly by breeders working either individually or collectively.
- Selection is focused partly on survival and partly on conformation and production traits.
- The gene pool is typically very broad at first, but often narrows dramatically due to population depletion.

- The populations tend to be extremely small.
- Geographic distribution is local.
- The primary advantage is exceptional adaptation to a specific environment.
- The primary drawback is extreme scarcity.

Feral Populations

A feral population is derived from the descendants of domestic animals that have escaped into the wild and survived to produce offspring. Subsequent generations are selected entirely by the environment, producing a breed that is remarkably well adapted to survival in a given geographic area.

The genetic base of a feral breed can be either very narrow or very broad, depending on how many founders there were and how much the population has been influenced by outside genetics. Thus, feral animals can be either highly inbred or highly diverse.

A feral population can be redomesticated and formed into either a landrace or a standardized breed. In fact, the Texas Longhorn breed went through all three phases.

Feral cattle breeds have not been included in this book, as they can be difficult to obtain and handle and are thus not likely choices for beginning cattle owners. The standardized White Park breed, however, has close relatives among feral cattle still living in England.

To summarize:

- Feral populations are developed primarily through natural selection.
- Multiple breeds from a variety of backgrounds typically form the genetic base.
- The selection process is guided by the environment.
- Selection is focused strictly on survival and reproduction.
- The gene pool may be either narrow or broad.
- The populations are typically very small.
- Geographic distribution is local or regional.

- The primary advantage is exceptional hardiness.
- The primary drawback is difficult temperament.

Country of Origin

Country of origin is another factor that separates one breed from another. This is because each country or world region has different cultural demands and environmental conditions.

The cattle breeds available in the United States today can be divided into five main groups based on geographical background:

- British.
- Continental.
- Zebu.
- American.
- Spanish-American.

British Breeds

America has long had an association with the British Isles, so it was only natural that British cattle breeds predominated on our shores for many years. The foundation of our British cattle population was imported beginning in the late 1700s. These importations continued well into the following century. Although British dairy breeds have declined in popularity since the Holstein dominated the scene, the vast majority of beef herds in America today are still built on British genetics.

Examples of British breeds include:

- Angus.
- Devon.
- Galloway.
- Guernsey.
- Hereford.
- Jersey.

- Red Poll.
- Shorthorn.

While each breed is slightly different, most British breeds share the following characteristics:

- Small size.
- Hardiness in cold climates.
- Early maturity.
- Fertility.
- Calving ease.
- High percentage of milk protein and butterfat.
- High percentage of waste at slaughter.
- Marbled beef.
- Tender meat.

British breeds have found niches in both commercial and alternative agriculture due to their adaptability. Although the beef breeds dominate the industry sale barns, they are also typically the breeds of choice for grassfed beef production. Likewise, the British dairy breeds still have a place in commercial milk production, but excel in low-input organic dairying. Several of the breeds, such as the Devon, fit into the dual-purpose category and can be used as all-around homestead cattle, providing beef, milk, and draft power for small farms.

Continental Breeds

The cattle of Continental Europe, many of them developed for draft purposes, first came to the attention of American cattlemen due to their impressive size. Although experiments were made with Continental breeds in the early 1900s, they did not become popular in the United States until the late 1960s and early 1970s, hence their other name—"exotic breeds."

These cattle were costly and difficult to obtain at first, so the process of establishing an American population was expedited by upgrading imports with British cattle already living on our shores.

Most Continental breeds were considered purebred after four or five generations of upgrading. They left their mark on the beef industry by increasing the frame size of the American cattle population, but this trend has abated somewhat in recent years along with the use of Continental genetics.

Examples of Continental breeds include:

- Belgian Blue.
- Charolais.
- Chianina.
- Gelbvieh.
- Holstein.
- Limousin.
- Maine Anjou.
- Piedmontese.
- Simmental.
- Tarentaise.

Continental breeds vary widely, but they tend to share a few traits:

- Large size.
- Late maturity.
- High milk yields.
- Rapid weight gain on feed.
- Large yield of beef.
- Low percentage of waste at slaughter.
- Lean beef.

While quite a few of the Continental breeds have potential as triple-purpose beef, dairy, and draft animals, they are rarely used in this way in America. One of the most important roles of Continental cattle in the United States is crossbreeding with British breeds to create hefty beef animals with good marbling. These crossbred cattle need plenty of grain to reach their full potential, so they are more commonly found in the feedlot than in a grassfed operation. The Holstein is the only major Continental breed in America used primarily for dairy purposes, and it has come to dominate the milk industry.

Zebu Breeds

All domestic cattle breeds belong to the species *Bos taurus*, but they are further divided into two subspecies:

- *Bos taurus taurus*, or taurine breeds, including the vast majority of breeds of European and American origin.
- *Bos taurus indicus*, or zebus; the humped cattle breeds.

Zebus are distinguished by big ears and baggy skin, features that help them dissipate heat easily in the hot, muggy climates from which they come. *Bos taurus indicus* breeds are often associated with India, but some of them have roots in other parts of southern Asia and in Africa.

The zebu is typically used for milk and draft purposes in its native home, two roles in which it has never excelled in America. While it is rarely eaten in many of its native countries, the zebu is elsewhere primarily used for raising crossbred beef steers. *Bos taurus indicus* has a reputation for being rather tough, but crossing it with a *Bos taurus taurus* animal can improve the meat quality of the offspring somewhat.

While there are hundreds of taurine breeds, there are relatively few zebu breeds. The only zebu breed familiar to most Americans is the Brahman. However, there are number of breeds in the United States that combine the influence of both subspecies.

In most parts of America, *Bos taurus taurus* is the preferred subspecies when it comes to cattle. The dairy breeds produce more milk than zebus, and the beef breeds produce tastier, more tender meat. For multipurpose cattle, there are many heritage breeds that better meet the requirements of homesteaders.

The arena in which *Bos taurus indicus* really shines is crossbreeding in extremely hot climates. Zebu breeds such as the

Brahman consistently impart heat, parasite, and fescue toxicosis resistance to their crossbred offspring, while the *Bos taurus taurus* influence improves beef and dairy performance to some extent. Many new and useful breeds have been developed specifically for the tropics in this way, such as the Beefmaster and the Santa Gertrudis.

Therefore, climate will determine whether a *Bos taurus taurus* or a *Bos taurus indicus* is right for you. For most parts of the United States, *Bos taurus taurus* is the animal of choice.

American Breeds

Besides the zebu-derived composite breeds, North America can boast of two unique groups of cattle breeds native to our hemisphere. The first group of breeds is highly reflective of the American melting pot.

In the early days of our country's settlement, people arrived from many different parts of Europe, often bringing their native breeds of cattle with them. Most of these, naturally, were various breeds of the British Isles:

- From Ireland, the Irish Moiled.
- From Scotland, the Ayrshire.
- From Wales, the Glamorgan.
- From England, the Shorthorn, the Hereford, and the English Longhorn.

In Canada, breeds from France were also introduced.

As the settlers mingled and adapted to their new homes, so did their livestock. Everywhere people farmed there were small populations of landrace cattle, especially adapted to their local conditions. These cattle were nearly always versatile, providing their owners with milk, meat, and draft power.

Unfortunately, multipurpose breeds fell out of favor after World War II. No one needed draft power anymore, and for beef and milk there were specialized cattle that could provide exceptional yields. Versatile landraces largely disappeared from the scene, preserved only on a few family farms in New England, the upper Midwest, and bordering areas of Canada. Several breeds even went extinct.

Some American breeds were rescued by the establishment of breed associations and the creation of breeding programs to raise production levels, particularly of milk. The Canadienne breed underwent such a process. Other breeds, such as the Lineback, were extensively crossbred to meet modern industrial demands, significantly altering their gene pools.

True American landraces are hard to find these days, but some feel that they are worth the effort due to their amazing adaptation to their traditional homes. A good example of the traditional American landrace is the Randall, a multipurpose breed representative of the early Lineback.

Spanish-American Breeds

The other group of breeds unique to the Americas descended from Spanish animals originally brought to Hispaniola by Christopher Columbus on his second voyage to the New World. From there, subsequent explorers took the cattle to the mainlands of North and South America in the 1500s.

The Spaniards frequently abandoned cattle during their travels, sometimes accidentally but sometimes with the express purpose of leaving breeding populations scattered in their wake to feed them on future journeys. The cattle were left to fend for themselves with the result that only the hardiest survived.

Americans often domesticated feral populations of Spanish cattle to take advantage of their low-maintenance beef production. Spanish cattle populations from different parts of North America developed along somewhat different lines due to both environmental factors and human selection. The result was distinct breeds. Compare, for instance, the Texas Longhorn, the Corriente, the Florida Cracker, and the Pineywoods.

Although selected for good disposition and beef characteristics in more recent years, the Spanish-American breeds still retain many traits from their feral background:

- Long horns.
- Highly variable coloration.
- Intelligence.
- Heat tolerance.
- Resistance to disease and parasites.
- Ability to thrive on poor-quality forage.
- Strong mothering instincts.
- Longevity.
- Very lean beef.

Despite their low-maintenance characteristics, the Spanish-American breeds are not mainstream beef breeds due to their horns, active natures, and lack of marbling. Most are also a little on the small side. However, the Spanish-American breeds have firmly established themselves as viable choices for grassfed beef operations.

The Pineywoods is unique among Spanish-American breeds in that it has potential for small-scale milk production. This breed and the Texas Longhorn also make good draft animals. The Corriente is the breed of choice for roping and other rodeo sports, although some cattle from the other Spanish-American breeds can perform satisfactorily in the arena, as well.

Heritage vs. Modern

Cattle breeds can be further classified as either modern or heritage breeds. Heritage breeds are breeds that were around before the days of industrial agriculture. In a sense, they are walking pieces of history.

In general, heritage breeds were adapted to very specific environments. Instead of being one-size-fits-all commodities, these animals developed distinctive traits that enabled them to thrive despite adverse local conditions, such as rough terrain, inclement weather, or poor-quality forage. Often they were raised with minimal supervision, which meant that they would be culled naturally and develop as low-maintenance landraces. Some, however, were standardized in an effort to improve their conformation and productive qualities. A few heritage breeds are feral populations that have been redomesticated.

Regardless of how they developed, heritage breeds were expected by their human owners to perform in spite of the difficulties they faced. Their tasks varied, but they had to thrive in their unique environments and still provide farmers with food and transportation. Often heritage cattle breeds were triple-purpose animals used for milk, meat, and draft work.

This long history of adaptation and hard work led to the development of many other useful characteristics:

- Structural soundness.
- Good immune systems.
- Ability to thrive in low-input systems.
- Intelligence.
- Longevity.
- Ability to reproduce with minimal intervention.

Today, many of these breeds still offer top-notch meat and milk, as well as distinctive looks not found in commodity breeds.

With so much going for them, why aren't heritage breeds more popular? There are two main reasons:

- Production.
- Uniformity.

Heritage breeds can rarely surpass their commercial counterparts in production. A Milking Devon will never produce as many gallons of milk as a Holstein. Heritage beef animals typically grow more slowly than modern breeds.

Furthermore, commodity agriculture is very standardized. Anything abnormal is typically penalized. For example, the steer that brings the most money at the sale barn is going to be hornless and solid black or red. A dun-colored Highland calf with a shaggy coat and the potential for long horns is going to sell at a heavy discount.

In short, only you can decide whether or not a heritage breed is a good fit for you. If quality food and low maintenance requirements are your aim, look no further. On the other hand, if you plan to market through more conventional channels, you may find that a modernized breed is a safer bet.

Where to Go From Here

The rest of this book will focus on the breeds themselves. While this resource should get you started in the right direction, there may be breeds of interest that have not been included. Also, if you see a breed that you like in this book, you will doubtless want to find more information.

Take a look at the "Helpful Resources" section at the end of the book for further reading material. Also be sure to search online for a breed organization for your breed of choice. While the websites of these organizations tend to be replete with one-sided promotional material, they do typically offer a wealth of useful information on the purchase, care, and breeding of the type of cattle in question.

CHAPTER 4
BEEF BREEDS

The beef breeds specialize in meat production. Historically, most of them descend from muscular draft stock that were repurposed in response to industrialization. Industrialization made mechanized agriculture possible, thus eliminating the need for draft animals. At the same time, however, it gave those animals new purpose by creating cities hungry for beef.

The specialization of the beef breeds started with Robert Bakewell in the 1700s. Working with the English Longhorn, he blazed a new trail for cattle breeders by seeking to make beef production more efficient. Specifically, Bakewell believed that heavy bones and high levels of milk production were a waste in cattle destined for slaughter. He selected cattle for traits such as overall fleshing ability and yield of high-value cuts of meat.

Bakewell's concept of breeding cattle exclusively for beef production remains somewhat novel around the world, primarily being practiced where extensive rangelands are available. North and South America are the leaders in specialized beef production due to their large-scale grassland ecosystems.

In the United States, most beef cattle are finished on grain in feedlots to maximize production and reduce the amount of time required to develop a quality carcass. Therefore, commercial cattle markets typically demand a large, fast-growing animal. In a grass-based system, cattle are typically selected for a smaller, more efficient size. They must also be chosen for the ability to retain good flavor and tenderness with fewer inputs.

Angus

Origin: Scotland, early 1800s.
Breed Structure: Standardized.
Population: Common.
Uses: Beef, crossbred beef.
Weight: Bulls average 2,200 lbs., cows average 1,450 lbs., but varies enormously; Lowline bulls 900–1,400 lbs., Lowline cows 700–1,000 lbs.
Preferred Climate: Cold to temperate; Red Angus can tolerate heat.
Temperament: Varies by bloodline; generally calm and good-natured, but some are unstable.
Health: Some bloodlines are extremely healthy; others have many genetic defects (dwarfism, extra legs, excessive bone formation, calves born dead due to fluid-filled cranial cavities, etc.).
Production: High dressing percentage, high ratio of top-dollar cuts, beef marbled but bland.

The Angus is America's top beef breed. A Certified Angus Beef program and a strong presence in the fast-food industry have made the Angus the choice of many in the cattle business. It is known for being hardy, adaptable, long-lived, and fertile, and it receives premium prices at the sale barn. However, popularity tends to attract incautious and sometimes unscrupulous breeders. Some bloodlines have produced cattle that are unstable in temperament and unsound in genetics.

Both black and red cattle were originally accepted for registration as purebred Angus until 1917, when the American Aberdeen Angus Association selected the more fashionable black as the color of choice and stopped registering reds. Some dedicated breeders, however, believed that the reds had all the advantages of the blacks, plus extra heat tolerance. Accordingly, the Red Angus Association of America was formed in 1954.

There is also a Lowline Angus developed in Australia in 1974 by selecting smaller animals from within the Angus breed. The project was initially an experiment to see what effect high and low growth rates had on herd profitability. The Lowline was found to produce quality beef remarkably efficiently. It is a uniformly docile animal that requires fewer inputs than its larger kin, and it is much easier on fences and pastures.

Barzona

Origin: United States, 1940s.
Breed Structure: Composite.
Population: Secure.
Uses: Beef, crossbred beef.
Weight: Bulls average 1,650 lbs., cows average 1,250 lbs.
Preferred Climate: Able to tolerate the harshest climates.
Temperament: Quiet, intelligent, strong herd instinct; can be dangerous if handled infrequently.
Health: No major problems.
Production: High beef yield, somewhat marbled.

This composite breed was developed in the mountains of Arizona to produce profitable quantities of beef while coping with a harsh and highly variable environment. The desired result was obtained after an impressive amount of research and years of planned matings, coupled with stringent culling so that only the most productive cows remained in the herd. The foundation animals were a blend of Angus, Hereford, Santa Gertrudis, and Africander (a heat- and tick-resistant mix of *Bos taurus taurus* and *Bos taurus indicus* hailing from South Africa).

Advantages of the Barzona include good disposition, insect resistance, disease resistance, longevity, fertility, easy calving, and a high-yielding, marbled carcass. This breed is also remarkable for its ability to maintain body condition on rough, brushy pastures.

Beefmaster

Origin: United States, 1930s.
Breed Structure: Composite.
Population: Common.
Uses: Beef.
Weight: Bulls over 2,000 lbs., cows 1,300–1,800 lbs.
Preferred Climate: Hot and dry, but able to tolerate some cold.
Temperament: Intelligent and gentle, but can be unpleasant if mistreated or unfamiliar with humans.
Health: No major problems.
Production: High yield, tends to receive low USDA grades.

This composite breed goes back to Tom Lasater's Texas-based herd, selected with unparalleled care for disposition, fertility, weight, conformation, hardiness, and milk production. The result was a blend of about 50% Brahman, 25% Shorthorn, and 25% Hereford. Today, it is believed to be one of the fastest-growing breeds in population numbers in America.

The Beefmaster is well adapted to formidably arid climates and is considered very profitable in such areas. Due to its zebu influence, it is also able to tolerate fescue pastures fairly well. In other environments, however, it tends to be docked at the sale barns for Brahman-related issues such as lower beef quality and the potential for temperament difficulties in herds that receive little interaction with humans.

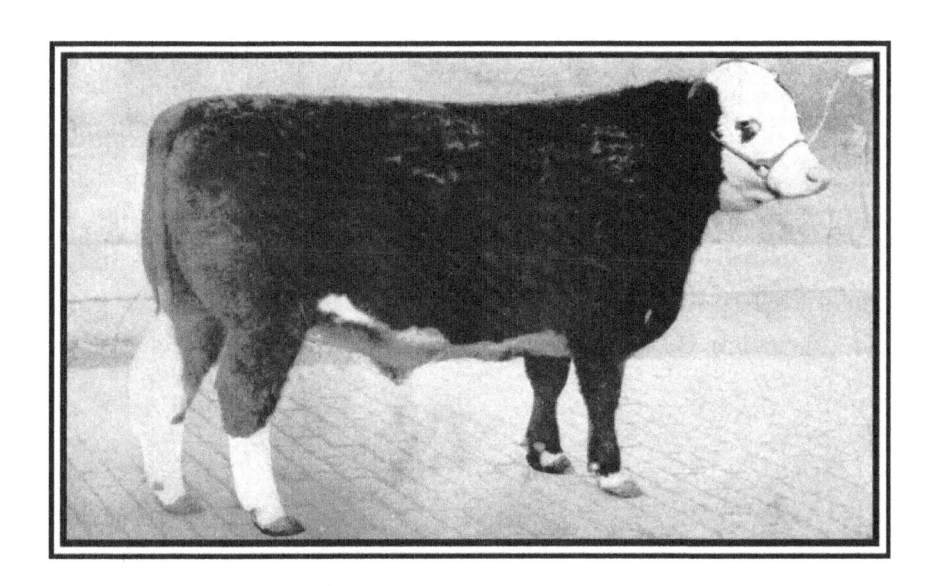

Black Baldy

Origin: N/A.
Breed Structure: Not a pure breed.
Population: Common.
Uses: Beef.
Weight: Varies.
Preferred Climate: Adaptable to most climates.
Temperament: Docile, good-natured.
Health: No major problems.
Production: Efficient beef producer, tendency toward fattiness, bland.

You've probably seen Black Baldies, the black cattle with the white faces. The Black Baldy, however, is not a true breed, but rather any combination of breeds that will produce the hallmark coloring.

White-faced black calves can be produced by crossing any black breed with any white-faced breed, such as Hereford or Simmental. Generally, though, Baldies are the offspring of an Angus x Hereford mating.

There is also a miniature Black Baldy produced by crossing a Lowline Angus with a miniature Hereford.

Hybrid vigor is the main reason for crossing two breeds of cattle, and it is a trait in which the Black Baldy excels. This cross has earned a reputation for being healthier than either purebred cattle or more nondescript "mongrels." It also has a good growth rate and is highly regarded at most sale barns. The Black Baldy qualifies for most premium beef programs, including Certified Angus Beef.

Brahman

Origin: United States, mid-1800s.
Breed Structure: Standardized.
Population: Common.
Uses: Crossbred beef, rodeo, crossbred dairy.
Weight: Bulls around 1,800 lbs., cows around 1,200 lbs.
Preferred Climate: Hot.
Temperament: Extremely intelligent; loyal when treated well, nervous or aggressive when handled roughly or infrequently; extremely protective of calves; a few can be unstable.
Health: Weak calf syndrome, allergic reactions to insecticides, skin and flesh injuries.
Production: Slow growth rate, lean, tough.

Few breeds are as well adapted to tropical heat, humidity, sunlight, parasites, insect-borne diseases, and low-quality forage as the Brahman. However, it is rarely kept outside of the South due to its drawbacks—a sometimes difficult personality, slow growth, failure to rebreed quickly after calving, and tough beef. For this reason, Brahman and Brahman-cross cattle often receive heavy discounts at sale barns in most parts of the United States.

While this zebu breed is typically used to produce crossbred beef in the American South, its relatives are primarily dairy and draft animals in other parts of the world. It can impart tremendous amounts of hybrid vigor and a good dose of heat tolerance to any crossbreeding system.

Brahmans and Brahman crosses tend to be resistant to fescue toxicosis, but they are not entirely immune.

Brangus

Origin: United States, 1930s.
Breed Structure: Composite.
Population: Common.
Uses: Beef, crossbred beef.
Weight: Bulls 1,800–2,000 lbs., cows 1,100–1,200 lbs.
Preferred Climate: Hot.
Temperament: Quiet, curious, and confident with familiar people; can be excitable if handled infrequently; some bloodlines aggressive and unpredictable.
Health: No major problems.
Production: Fast growth, high yield.

The Brangus is yet another Brahman-derived breed developed for the challenging conditions of hot, humid climates. The precise blend in this composite is 5/8 Angus and 3/8 Brahman. Both the Black Brangus and the Red Brangus are becoming increasingly popular in the South.

The Brangus is hardy and resistant to many diseases and health problems, particularly bloat. Both color varieties excel in resistance to hot-weather difficulties such as pinkeye and sunburn, but the Red Brangus is sometimes considered to have a particular advantage under pressure from extreme temperatures.

Longevity, calving ease, and adaptability to both forage- and grain-based production systems give the Brangus a great deal of appeal in its native climate. Outside of the South, however, it may suffer from the cold and will typically be docked at the sale barn for its Brahman influence and associated temperament and fertility challenges.

British White

Origin: Great Britain, ancient.
Breed Structure: Standardized.
Population: Secure.
Uses: Beef, crossbred beef, crossbred dairy, conservation grazing.
Weight: Bulls about 1,850 lbs., cows about 1,250 lbs.
Preferred Climate: Temperate, but tolerant of extremes.
Temperament: Extremely calm and docile, intelligent; marked ability to distinguish between real and perceived threats to calves.
Health: No major problems.
Production: Very high yields, well marbled, tender and succulent with excellent flavor.

Most farmers and ranchers who have worked with British Whites swear by them for the rest of their lives. This breed is unsurpassed for good temperament and is able to maintain good health throughout a long productive lifespan. It is also remarkably well suited to grassfed and organic beef production due to its ability to thrive on rough, brushy, or marshy pasture alone while raising healthy calves unassisted and producing gourmet meat.

Even though the hair of this breed is white, sunburn and pinkeye are not issues with the British White because its exposed skin is pigmented either black or red. This trait is useful in areas with hot summers, but is nevertheless penalized at the sale barn.

Charolais

Origin: France, ancient.
Breed Structure: Standardized.
Population: Common.
Uses: Crossbred beef.
Weight: Bulls average 2,400 lbs., cows average 1,800 lbs.
Preferred Climate: Adaptable to most climates.
Temperament: Varies between skittish and aggressive; a few bloodlines are good-natured.
Health: Prolapse, calving problems, sunburn, pinkeye, cancer eye.
Production: Abundance of high-value cuts, lean, tender, good flavor.

This Continental breed is typically crossbred with British cattle to take advantage of its genetics for growth and abundant lean meat production. Charolais-influenced crossbred calves typically perform well at the sale barn. Note that they require lush pastures and substantial feeding to reach their full potential.

Neither the Charolais nor Charolais-cross cattle are recommended for beginners due to their problematic dispositions and their tendency toward difficult births.

Chianina

Origin: Italy, ancient.
Breed Structure: Standardized.
Population: Common.
Uses: Crossbred beef, draft.
Weight: Bulls average 2,850 lbs., cows 1,750 lbs.
Preferred Climate: Adaptable to most climates.
Temperament: Varies between nervous and aggressive, but generally unstable; a few bloodlines are docile.
Health: No major problems.
Production: Outstanding carcass weights, high beef yields, lean, good taste.

The Chianina is the largest breed of cattle in the world. This is why it has been a popular choice for crossbreeding beef cattle in America. Large size translates into positive production traits such as heavy carcass weight and ample yields of beef. Unfortunately, it also translates to a hefty appetite.

This is a hardy breed overall. Even though its coat is white, pigmented skin protects it from sunburn and pinkeye. It displays some resistance to parasites. The Chianina is far less prone to calving difficulties than many large breeds.

Florida Cracker

Origin: United States, 1500s.
Breed Structure: Landrace.
Population: Endangered.
Uses: Beef, roping.
Weight: Bulls 800–1,200 lbs., cows 600–800 lbs.
Preferred Climate: Hot and humid.
Temperament: Docile, active; varies by bloodline.
Health: Dwarfism.
Production: Lean, flavorful.

The rare Florida Cracker is descended from Spanish cattle and has adapted over several centuries to rustle its own living in the hot, humid, buggy Florida scrub. In its native home, these characteristics result in cows that are reproductively efficient and steers that excel at low-input beef production.

The Florida Cracker is a small breed. A few cattle are even miniatures, known as "Guineas." While Guineas are extremely low-maintenance cattle, they should not be bred together or the resulting calf may suffer from lethal dwarfism.

In appearance, the Florida Cracker is much like its close relative, the Texas Longhorn, sporting a wide range of flashy colors and long, twisted horns (albeit not quite as long as the standardized Longhorn).

This breed is also of great historical importance to its home state, being closely associated with the cracker cowboy culture of the 1800s and early 1900s. The crackers were Florida cowboys who made a living searching for free-roaming cattle in the scrub and driving them to ports on the Gulf of Mexico. They received their name from the practice of cracking cow whips to drive the herds.

Galloway

Origin: Scotland, ancient.
Breed Structure: Standardized.
Population: Rare.
Uses: Beef, hides, petting zoos, conservation grazing, guarding sheep.
Weight: Bulls 1,300–2,000 lbs., cows 1,000–1,300 lbs.
Preferred Climate: Cold.
Temperament: Gentle and friendly toward humans; wary, protective of the herd, aggressive toward dogs; some cows may bully calves other than their own.
Health: Heat stress; udder problems and incomplete lung development in belted variety.
Production: Good yields of high-value cuts, lean, exceptionally tender, flavorful, juicy.

For areas with cold winters, the Galloway is an excellent beef breed due to its good nature, its easy-keeping tendencies, and its ability to calve unassisted. It can even thrive at high altitudes. The cows of this breed may continue to calve annually up to 20 years of age. It consistently produces high-quality beef when finished on pasture—even weedy pasture—provided it is allowed enough time to grow. Black Galloways also receive good prices at the sale barn.

The popular Belted Galloway is rather different than the other varieties, as a few individuals may be suitable for home dairy purposes. However, it is frequently bred as a novelty and has subsequently suffered in health and soundness.

Gelbvieh

Origin: Germany, early 1900s.
Breed Structure: Standardized.
Population: Common.
Uses: Crossbred beef.
Weight: Bulls average 2,300 lbs., cows average 1,400 lbs.
Preferred Climate: Temperate to hot.
Temperament: Docile if handled properly; some individuals wary.
Health: Udder problems, calving problems in some individuals.
Production: Moderate yield, lean.

As is the case with many Continental beef breeds, the Gelbvieh grows quickly, but requires plenty of feed to compensate. This breed does have better fertility than many big, fast-growing breeds, however.

Note that the Gelbvieh can be rather variable in many characteristics. It is best to seek breeding stock from a reputable producer who is evaluating and selecting cattle for traits of importance, such as disposition and calving ease.

Hereford

Origin: Great Britain, early 1600s.
Breed Structure: Standardized.
Population: Common.
Uses: Crossbred beef, beef, draft.
Weight: Bulls 2,200–2,650 lbs., cows 1,300–1,750 lbs., but varies widely; miniature bulls about 1,000 lbs., miniature cows 650–800 lbs.
Preferred Climate: Adaptable to most climates.
Temperament: Easygoing, affectionate, extremely calm.
Health: Sunburn; some bloodlines have numerous health issues such as prolapse and arthritis.
Production: High yield (particularly of valuable cuts), tasty, tender.

The Hereford is a common choice for a grass-based beef business, as it grows quickly while being hardy and low-maintenance, not to mention easy to get along with. Quality Herefords are also known for fertility, calving ease, and longevity. Some bloodlines have even been selected for high-altitude adaptation, as well.

Herefords come in several varieties. The classic red Hereford is currently the second most popular breed in the nation, partly because of its proven versatility in crossbreeding, partly because of its own desirable characteristics, such as docility and high-quality beef. The newer black Hereford is also increasing in numbers on commercial operations, as it is useful in breeding Black Baldies, while miniature Herefords are becoming popular for small acreages.

Be aware that some Herefords have been bred primarily for large size. This has come at the expense of health, soundness, and calving ease. Before purchasing a Hereford, check to make sure the breeder's philosophies align with your own, and also note that truly good breeding animals typically come with a substantial price tag.

Limousin

Origin: France, ancient.
Breed Structure: Standardized.
Population: Common.
Uses: Crossbred beef, beef.
Weight: Bulls average 2,400 lbs., cows average 1,300 lbs.
Preferred Climate: Cold, but able to tolerate most climates.
Temperament: Calm, good-natured; a few can be unpredictable.
Health: No major problems.
Production: High yield of top-dollar cuts, lean, tender, good flavor.

Limousin breeders have decidedly changed the direction of their breed in the last few decades. In the early 1990s, restoring docility to the Limousin was chosen as the number-one goal. Since then, dedicated cattlemen have been tackling one weakness after another, shaping the Limousin into a more practical beef breed. Tap into this work by purchasing cattle from a reputable source.

While the Limousin does have the late maturity and high nutritional requirements characteristic of most Continental beef breeds, it excels in the areas of longevity, fertility, calving ease, and beef quality.

Maine Anjou

Origin: France, mid-1800s.
Breed Structure: Standardized.
Population: Common.
Uses: Beef, crossbred dairy.
Weight: Bulls average 2,800 lbs., cows average 2,000 lbs.
Preferred Climate: Cold to temperate.
Temperament: Gentle but stubborn.
Health: Calving problems.
Production: High yield, lean, tender.

Originally a dual-purpose breed, the Maine Anjou is now primarily valued for its adaptation to the feedlot, growing rapidly on feed to produce a lean, muscular carcass. This rapid growth has come at the expense of calving ease, however.

The history of the Maine Anjou is interesting, as it is a prime example of a composite breed that has become standardized over time. It was originally 3/4 Shorthorn and 1/4 Mancelle, a large French breed known for fattening easily.

Santa Gertrudis

Origin: United States, early 1900s.
Breed Structure: Composite.
Population: Common.
Uses: Beef, crossbred beef.
Weight: Bulls 1,700–2,200 lbs., cows 1,350–1,850 lbs.
Preferred Climate: Hot, but can tolerate most climates.
Temperament: Generally mild and docile, but may be slightly nervous; bulls can be aggressive.
Health: No major problems.
Production: Lean beef, high grades when allowed longer to finish.

This breed is a Texas-developed blend of 5/8 Shorthorn and 3/8 Brahman. While the predominance of British blood in this composite provides a quality carcass, the Brahman influence contributes resistance to heat and insects.

Other desirable traits of the Santa Gertrudis include its fertility, calving ease, and ability to gain weight on pasture alone. It has a high degree of resistance to bloat.

Texas Longhorn

Origin: United States, early 1500s.
Breed Structure: Standardized.
Population: Common.
Uses: Beef, crossbred beef, conservation grazing, draft, novelty, horns, hides, roping.
Weight: Bulls 1,400–2,200 lbs., cows 600–1,400 lbs.
Preferred Climate: Able to tolerate most harsh climates.
Temperament: Docile, friendly, highly intelligent; can be wild if handled infrequently.
Health: Broken horns.
Production: Lower beef yields, extremely lean, good flavor.

In recent years, a rising demand for lean beef has created an ideal niche for the Texas Longhorn. This breed is extremely well suited to grassfed operations, being able to thrive on poor pasture with few parasite or disease problems. It is also quite capable of breeding and calving unassisted, often for 20 to 30 years.

Popularity may eventually prove to be the breed's downfall. Many Texas Longhorn breeders emphasize raising cattle for the "total package," which in this case means color, conformation, and horn length. To accomplish this purpose, longhorn ranchers have bred extensively to only one of the original bloodlines—the one considered best for show purposes. The other bloodlines are approaching extinction as unique identities, having been blended together to improve show qualities.

So far no major genetic problems have come to light. The Texas Longhorn is still hardy and fertile, and it also has much better manners than it did in the days of the cattle drives. But a few longtime breeders are starting to feel that the good old Texas Longhorn "ain't what it used to be." Time alone will reveal the ultimate fate of this magnificent breed.

CHAPTER 5
DAIRY BREEDS

Ancient peoples have used milk, butter, and other dairy products since the earliest times, as testified by artwork from the temples and tombs of Mesopotamia and Egypt, as well as the Sanskrit writings of India. In fact, in some societies, dairying was the most important purpose of cattle; draft work was a way to gain use from surplus animals, while beef was strictly for ceremonies and special occasions.

The specialization of dairy breeds, however, is a more recent phenomenon. Cattle breeds selected strictly for dairy production first began to appear in the early 1800s, starting in northern Europe. This process was brought about by the Industrial Revolution and the subsequent rise of hungry urban populations. Specialization was further spurred on by growing export markets around the world, as well as by the rise of factories to process dairy products into cheese and butter in the mid-1800s.

The most common dairy breed worldwide is the Holstein, an American offshoot of the Friesian breed from the Netherlands. Since the industrial Holstein was developed, it has largely replaced all other breeds in commercial dairying. Many of the surviving dairy breeds have been upgraded with Holstein genetics in an effort to retain their competitive advantage.

The Holstein has dominated the world of dairying because it can produce more milk than any other breed. This increased production comes at the cost of quality, however. Holstein milk contains a high percentage of water and a form of beta casein difficult for many people to digest.

Surplus dairy calves are frequently raised for beef, but there are still important differences between dual-purpose cattle and true dairy cattle. Dairy cattle are almost exclusively selected for dairy qualities, rarely if ever for beef qualities. Most dairy calves finished for beef are actually the result of crossing the dairy cow with a beef sire. Dual-purpose breeds are selected for versatility or have separate bloodlines specialized for each purpose.

It is important to evaluate whether a high-producing dairy breed or a versatile dual-purpose breed better meets your needs. As a general rule, dairy breeds are better for a commercial or agripreneurial situation, while dual-purpose breeds may only be able to supply the family refrigerator with milk and butter. However, some specialized dairy bloodlines of dual-purpose breeds can fit into small farm businesses depending on the quantities of milk required.

There are fewer dairy breeds than beef breeds to choose from. If you decide that a specialized dairy breed is right for you, the next step is to assess your options and decide which breed best fits your environment, management system, and production needs.

Note that total milk production is typically inversely related to protein and butterfat percentage. A breed that excels in producing large quantities of fluid milk is generally not the best choice for a business specializing in value-added dairy products such as cheese. Likewise, a breed that produces high percentages of protein and butterfat typically does so at the expense of total milk production.

Ayrshire

Origin: Scotland, mid-1700s.
Breed Structure: Standardized.
Population: Rare.
Uses: Dairy, draft, beef (mostly crossbred).
Weight: Bulls about 2,000 lbs., cows about 1,400 lbs.
Preferred Climate: Adaptable to most climates.
Temperament: Friendly, mild, smart, spunky, slightly stubborn.
Health: No major problems.
Production: Average milk yield 17,000 lbs. annually, 3.3% protein, 3.9% butterfat.

Because of its efficiency in grass-based dairying, the Ayrshire is becoming increasingly popular with organic farmers. The high levels of butterfat in its milk make it an excellent choice for value-added dairy products, such as butter, cheese, yogurt, and ice cream. Even better, the fat globules in this breed's milk are small and easy to digest. However, prospective buyers will want to investigate their Ayrshire genetics carefully, as some cows carry a gene that causes their milk to taste like fish.

Overall, this is a low-maintenance dairy breed known for soundness, calving ease, and a long working lifespan. An Ayrshire heifer will typically calve for the first time at around 29-1/2 months of age. Its calves are typically born hale and hearty. The cows have excellent foraging instincts and do not require perfect pasture to maintain their production levels.

Canadienne

Origin: Canada, early 1600s.
Breed Structure: Standardized.
Population: Endangered.
Uses: Dairy.
Weight: Bulls average 1,600 lbs., cows average 1,100 lbs.
Preferred Climate: Cold.
Temperament: Very docile.
Health: No major problems.
Production: Average milk yield 8,000 lbs. annually, 3.6% protein, 4.4% butterfat.

While not a high-producing breed, what the Canadienne does produce, it produces with few inputs. It is extremely hardy and quite skilled at foraging for its own living.

The milk of this breed is well suited to cheese production. While it has potential for lean beef production, this purpose is not recommended at present due to the breed's extreme scarcity.

Dutch Belted

Origin: Netherlands, 1600s.
Breed Structure: Standardized.
Population: Endangered.
Uses: Dairy, novelty, draft, crossbred beef.
Weight: Bulls up to 2,000 lbs., cows 900–1,500 lbs.
Preferred Climate: Temperate.
Temperament: Sweet, trusting, patient, intelligent.
Health: Sensitivity to insecticides.
Production: Average milk yield 13,000 lb. annually, 3.0% protein, 3.6% butterfat.

While some hobby farmers keep Dutch Belted cattle just for their looks, the primary purpose of the breed is to produce milk, particularly in grass-based systems. Dutch Belted milk works well for making cream, butter, and cheese, but it is exceptionally delicious just as a beverage. The milk has an excellent flavor, high levels of beta carotene, and small fat globules that are easy on the digestive system.

This is a hardy breed with strong foraging instincts and the ability to maintain body condition on limited resources. It calves easily and consistently up to 20 years of age. The calves are born vigorous.

Guernsey

Origin: Channel Islands, mid-900s.
Breed Structure: Standardized.
Population: Rare.
Uses: Dairy, crossbred dairy, crossbred beef, draft.
Weight: Bulls average 1,650 lbs., cows average 1,100 lbs.
Preferred Climate: Adaptable to most climates.
Temperament: Gentle, cheerful, quiet, intelligent; bulls are aggressive.
Health: Lung problems, udder infections.
Production: Average milk yield 16,000 lbs. annually, 3.5% protein, 4.3% butterfat.

Although the Guernsey is still one of America's favorite dairy breeds, its fortunes have not recovered since the Holstein phenomenon began. In an effort to compete, some breeders are upgrading their Guernsey herds with the red-and-white variety of Holstein to increase production. It would appear that the Guernsey is currently caught between two conflicting trends: increasing milk yields to meet the needs of commercial mass production and preserving traditional characteristics to fill the demands of an increasing number of small farms. How the conflict will be resolved in this breed remains to be seen.

Meanwhile, inbreeding increases as the breed's numbers decline. This has led to some delicate constitutions among Guernsey cows. Seeking out less inbred milk cows is worthwhile, as when healthy this breed excels at rustling its own living on low-quality pasture.

Owners should be aware that the Guernsey reaches puberty at a young age. Heifers should not be bred too early, or they may injure themselves producing more milk than they can support. They can safely have their first calf at around age two. The calves tend to be frail.

The composition of Guernsey milk is ideal for making cream, butter, cheese, and yogurt. Interestingly, it is also just right for making a stable foam in specialty coffees. Many Guernseys have the gene for easy-to-digest A2 beta casein.

Holstein

Origin: United States, mid-1800s.
Breed Structure: Industrial.
Population: Common.
Uses: Dairy, crossbred dairy, crossbred beef, draft.
Weight: Bulls average 2,200 lbs., cows average 1,500 lbs.
Preferred Climate: Cold to temperate.
Temperament: Generally calm, agreeable; bulls are extremely vicious.
Health: Many health problems, ranging from lameness to metabolic disorders.
Production: Average milk yield 25,000 lbs. annually, 3.2% protein, 3.7% butterfat.

While the Holstein, or Holstein-Friesian, is typically considered a Dutch breed, the modern Holstein is far removed from the original stock. The breed as we now know it is derived from a small portion of the Dutch Friesian population imported to America. This American population was isolated in the mid-1800s when an outbreak of foot-and-mouth disease in the Netherlands ended further importations of Dutch cattle. Americans then selected their cattle for larger frame sizes and higher milk yields. By the 1960s, the American Holstein had emerged as the most influential dairy breed worldwide.

The reason that the Holstein is so popular is its incredible milk yield. It also matures quickly and has an udder shape that is easy to milk, both important dairy traits. However, high production levels come at the cost of extremely high maintenance requirements, low quantities of protein and butterfat in the milk, and a very short lifespan. This is not a breed that can be turned out on any old pasture and expected to thrive—milking Holsteins require top-notch forage and plenty of it. Most require supplemental grain, as well.

Jersey

Origin: Channel Islands, mid-900s.
Breed Structure: Standardized.
Population: Common.
Uses: Dairy, draft, crossbred beef, crossbred dairy.
Weight: Bulls 1,200–1,800 lbs., cows 800–1,200 lbs.; miniature bulls average 850 lbs., miniature cows up to 700 lbs.
Preferred Climate: Temperate to hot.
Temperament: Highly sensitive, but very affectionate if handled gently; bulls are extremely vicious.
Health: Metabolic disorders, pinkeye; calves are prone to hypothermia and dehydration.
Production: Average milk yield 17,000 lbs. annually (8,000 lbs. annually for miniatures), 3.8% protein, 4.8% butterfat.

The Jersey is the second most popular dairy breed in America, and the fastest-growing dairy breed in numbers worldwide. A miniature variety suited to homesteading exists and is also enjoying great popularity.

In an effort to make this breed competitive with the Holstein, breeders have emphasized the size and total milk production of their cows at the expense of protein and butterfat. Even so, Jersey milk is still extremely rich, packed with vitamins and minerals, and well suited to making cheese, butter, yogurt, and ice cream.

The Jersey is a favorite with organic dairies. It has strong grazing instincts and is still small enough to be low-impact in the pasture. However, it is rather delicate. Heifers should be allowed to mature a bit before calving for the first time, and careful tending is required to prevent high mortality rates among newborn calves.

Kerry

Origin: Ireland, ancient.
Breed Structure: Standardized.
Population: Endangered.
Uses: Dairy.
Weight: Bulls up to 1,200 lbs., cows up to 850 lbs.
Preferred Climate: Temperate.
Temperament: Mannerly but spirited, intelligent.
Health: No major problems.
Production: Average milk yield 6,000 lbs. annually, 3.2% protein, 4.0% butterfat.

This diminutive cow is a good choice for smaller acreages where a lower milk yield is acceptable. In fact, this has likely made the Kerry more popular in America now than at any other time in its history. While the Kerry may not be able to compete with larger dairy cows for total production, it compensates with the ability to produce high-quality, easily digestible milk on sparse pasture with little or no supplemental feeding.

The Kerry is considered a breed best for slightly more experienced cattle owners because of its temperament. While it has a good disposition overall, it abounds in nervous energy.

Lineback

Origin: United States, 1700s.
Breed Structure: Standardized.
Population: Secure.
Uses: Novelty, dairy.
Weight: Bulls average 1,800 lbs., cows average 1,200 lbs.
Preferred Climate: Adaptable to most climates.
Temperament: Varies between docile and energetically stubborn.
Health: Calving problems in some bloodlines.
Production: Average milk yield 23,000 lbs. annually, good protein, good butterfat.

The Lineback was originally a landrace breed, derived from a little bit of anything adapted to providing meat, milk, and draft power in early America. After World War II, however, the Lineback was upgraded extensively with Holstein blood to increase its milk production. Until very recently, registration was based on appearance, not parentage. Most Linebacks now probably have more Holstein than landrace in them.

Despite the Holstein influence, the Lineback remains an

adaptable, long-lived breed able to forage for its own living. Note that it is highly variable in many characteristics, so prospective owners should take care to seek individual cattle suited to their needs.

CHAPTER 6
DUAL-PURPOSE BREEDS

Dual- and triple-purpose breeds were once the norm across much of the world. From the earliest times up to the advent of industrialization, cows produced milk, steers provided draft power, and old cattle of both sexes provided beef at the end of their working lifespans.

The Industrial Revolution changed this. For one thing, it led to the rise of urban population centers with a high demand for plenty of food to nourish the working classes. For another thing, it eventually led to the mechanization of agriculture. Most cattle breeds were repurposed to specialize in either beef or dairy. Some interest in less specialized breeds has returned with the resurgence of sustainable agriculture.

Today, dual-purpose breeds come in several flavors. Some are actually triple-purpose cattle equally suited for milk, meat, and draft purposes. These breeds are typically excellent all-around homestead cattle.

Quite a few dual-purpose breeds come in two varieties—one type specialized for beef production and one type specialized for dairy production, with very little intermingling of the two types.

A few breeds contain a specialized type and a generalized type. One example would be the Devon, which includes the specialized Beef Devon, or Red Devon, and the American Milking Devon—an all-around triple-purpose animal.

And, finally, it is worth mentioning that many cattle breeds are still considered dual- or triple-purpose in their native countries but have become specialized upon their arrival in the United States. When the American populations of these breeds still contain the genetic potential for satisfactory all-around performance, they have been included in this chapter. Examples include the Red Poll (typically used for beef in America but with plenty of potential as a homestead milk cow) and the Brown Swiss (often considered a dairy breed, but one that also makes an excellent beef animal).

Brown Swiss

Origin: Switzerland, ancient.
Breed Structure: Standardized.
Population: Common.
Uses: Dairy, beef, draft, crossbred dairy.
Weight: Bulls average 2,200 lbs., cows average 1,350 lbs.
Preferred Climate: Adaptable to most climates.
Temperament: Gentle, quiet, and good-natured, but very stubborn.
Health: Arachnomelia syndrome (a skeletal malformation also known as "spider legs").
Production: Average milk yield 20,000 lbs. annually, 3.5% protein, 3.9% butterfat; lean beef.

This old breed has all the makings of a good dual-purpose cow, although its milk production may be a little high for homesteading purposes. It is healthy, durable, and insect-resistant. It can handle most environments with ease and will continue to milk until 12 to 15 years of age. Be aware that the breed has increased in size over time, however, resulting in an enhanced appetite.

The Brown Swiss heifer usually calves for the first time at about

28-1/2 months of age. Calves are born easily, but can be difficult to rear. They go down quickly under adverse conditions and are nearly impossible to train to nurse from a bottle.

Brown Swiss milk is ideally suited to high-quality cheese production. The milk is also quite delicious when made into butter or yogurt, or simply as a beverage. This breed produces satisfactory grassfed beef, albeit rather slowly.

Devon

Origin: Great Britain, ancient.
Breed Structure: Standardized.
Population: Beef type rare; milking type endangered.
Uses: Beef, conservation grazing; crossbred beef (beef type); dairy, draft (milking type).
Weight: Bulls 1,600–2,000 lbs., cows average 1,100 lbs.
Preferred Climate: Adaptable to most climates.
Temperament: Docile, friendly.
Health: No major problems; calving problems in larger beef-type cattle.
Production: Fine-flavored beef, juicy, tender; average milk yield 4,300 lbs. annually, 3.6% protein, 4.2% butterfat.

The Devon comes in two varieties—the specialized Beef or Red Devon and the triple-purpose American Milking Devon. Both versions are hardy, adaptable, low-maintenance cattle that thrive in grass-based production systems with few health problems. Whether raised for meat or milk, this breed is guaranteed to deliver a gourmet product every time.

There are some temperament differences between the beef type and the milking type. In general, the beef type is somewhat less intelligent, but calmer and easier to handle. The milking type is smarter, livelier, and pluckier.

Both types of Devon have earned an excellent reputation for their good health. They seem to have a natural resistance to many diseases. Sun-related problems like cancer eye are very rare in Devons, and lameness is virtually unknown. Bloat is also rare in this breed.

Dexter

Origin: Ireland, late 1700s.
Breed Structure: Standardized.
Population: Rare.
Uses: Dairy, beef, crossbred beef, draft, novelty, crossbreeding to create miniature cattle breeds.
Weight: 700–1,000 lbs.
Preferred Climate: Temperate.
Temperament: Docile but spirited, wary if handled infrequently, very vocal.
Health: Dwarfism.
Production: Average milk yield 7,000 lbs. annually, 3.5% protein, 4.0% butterfat; low beef yield, fine-textured, tender, very tasty.

This miniature breed produces milk and meat in quantities too small for commercial purposes, but what it does produce it produces very efficiently. It can thrive on weeds and brush if necessary, it winters well with relatively little hay, and it doesn't destroy pastures with its sheer mass.

Other qualities of the remarkable Dexter are its ability as a mother and its strength as a draft ox—quite impressive for its size. Note, however, that careful attention must be paid to selecting good breeding stock, as dwarfism is a problem in this breed.

Highland

Origin: Scotland, ancient.
Breed Structure: Standardized.
Population: Rare.
Uses: Novelty, beef, crossbred beef, dairy, draft, horns, fiber, hides.
Weight: Bulls 1,300–1,750 lbs., cows 950–1,150 lbs.
Preferred Climate: Cold.
Temperament: Confident, docile, intelligent, strong herd instinct; can be wild if handled infrequently.
Health: Heat stroke, overgrown hooves; leg and udder problems in commercial bloodlines.
Production: Good beef yield, very lean, rich taste much like bison, fine texture; average milk yield 2,600 lbs. annually, 3.6% protein, 4.6% butterfat.

In America, the Highland has a reputation as a novelty animal kept mainly for its good looks. Although it does make a good pet, tourist attraction, or photographer's model, the Highland has so much more to offer.

This breed is remarkably hardy and self-sufficient, able to thrive on the poorest pastures. It is also capable of reproducing without intervention and will frequently continue to calve annually until 15 to 20 years of age. The Highland does require some attention to grooming and hoof trimming for best health. It is prone to ticks and lice in hot climates. Its beef is lean but flavorful, providing a unique, quality product when cooked gently.

Unfortunately, the Highland breed is currently caught in a tug-of-war between the conflicting interests of the show ring and the commodity market. Some bloodlines are raised mainly for looks, while others are bred to be fast-growing producers of conventional marbled beef. What will become of the traditional hardy Highland remains to be seen.

Also note that, while good-natured, Highlands are not lazy pushovers! They are incredibly intelligent and athletic. If they tire of their present situation, they will make a quick escape.

Normande

Origin: France, 800s.
Breed Structure: Standardized.
Population: Secure.
Uses: Dairy, beef, crossbred beef, crossbred dairy, draft, novelty.
Weight: Bulls average 2,200 lbs., cows average 1,450 lbs.
Preferred Climate: Adaptable to most climates.
Temperament: Calm, gentle, slow-moving; can be shy if unfamiliar with humans.
Health: No major problems.
Production: Average milk yield 14,000 lbs. annually, 3.5% protein, 4.2% butterfat; high yield of top-dollar beef cuts, lean, tender, excellent flavor.

The Normande has proven to be a great choice for forage-based systems—no matter what it is required to produce. On pasture, it is a healthy, vigorous breed that can deliver gourmet food every time. Confinement and high-energy feed will lead to obesity, lameness, metabolic disorders, and calving difficulties.

Randall

Origin: United States, early 1900s.
Breed Structure: Landrace.
Population: Endangered.
Uses: Dairy, draft, beef.
Weight: 1,100–1,600 lbs.
Preferred Climate: Cold.
Temperament: Very intelligent, wary, remarkably affectionate toward people they trust; bulls vary from calm to unpredictable.
Health: No major problems.
Production: Average milk yield 5,000 lbs. annually, 3.2% protein, 3.7% butterfat; tender beef, very flavorful.

The Randall represents the original American Lineback prior to the introduction of Holstein blood. It is derived from a single herd owned by the Randall family of Vermont. The rescue of this breed from a population of nine females and six males is one of the most dramatic stories in livestock breed conservation history. A careful breeding plan has saved the Randall from inbreeding problems and has resulted in a diverse gene pool with cattle well adapted to nearly every conceivable purpose.

This breed displays many natural instincts lost in more modern breeds, such as self-reliance and mothering ability, and it has superb health and hardiness. Both its meat and its milk are noted for quality and flavor.

Red Poll

Origin: Great Britain, early 1800s.
Breed Structure: Standardized.
Population: Endangered.
Uses: Beef, crossbred beef, dairy.
Weight: Bulls average 1,800 lbs., cows average 1,200 lbs.
Preferred Climate: Cold to temperate.
Temperament: Quiet, docile, curious.
Health: No major problems.
Production: High beef yield, excellent flavor, tender; average milk yield 9,000 lbs. annually, 3.6% protein, 3.8% butterfat.

The Red Poll is a highly adaptable breed that thrives in many beef production systems. It excels, however, in providing low-input beef for marketing to health-conscious customers and those seeking gourmet meat. More conventional beef operations can also benefit from using Red Polls in crossbreeding programs to achieve docile calves with quality carcass traits.

Dairying with Red Polls is not very common anymore, but it can be done on a small scale. This versatility makes the breed a good choice for homesteaders seeking dual-purpose cattle.

Regardless of purpose, most cattle owners can appreciate the excellent health and temperament of this breed, not to mention its ability to thrive on poor-quality forage. The Red Poll is very fertile and makes a good mother, but note that it may have calving difficulties if allowed to calve before three years of age.

Shorthorn

Origin: Great Britain, 1700s.
Breed Structure: Standardized.
Population: Beef type common, milking type endangered.
Uses: Beef; crossbred beef (beef type); dairy, draft (milking type).
Weight: Beef bulls 1,800–2,200 lbs., cows average 1,800 lbs.; milking bulls average 2,000 lbs., cows 1,200–1,400 lbs.
Preferred Climate: Cold to temperate.
Temperament: Slow-moving, mild, trainable.
Health: Genetic skeletal malformations, incomplete lung development, sterility, pinkeye, chronic intestinal infection.
Production: Exceptionally tender beef, rich flavor, unusual texture; average milk yield 17,000 lbs. annually, 3.4% protein, 3.8% butterfat.

Modern Shorthorns are rather different from their ancestors. The Beef Shorthorn has been selected for rapid growth, while the Milking Shorthorn has been extensively upgraded with Holstein genetics. Genetic defects are rampant in both types, while hardiness has suffered.

Old-fashioned triple-purpose bloodlines, while scarce, do exist. The original heritage Shorthorn remains hardy, self-sufficient, and fertile, a breed still worth seeking out for low-input farming.

In addition to the health issues listed above, Shorthorns may suffer from obesity and lameness when fed grain. This breed tends to perform poorly on fescue pastures.

Simmental

Origin: Switzerland, Middle Ages.
Breed Structure: Standardized.
Population: Common.
Uses: Crossbred beef, dairy, draft.
Weight: Bulls average 2,400 lbs., cows average 1,650 lbs.
Preferred Climate: Adaptable to most climates.
Temperament: Docile, relaxed.
Health: Mastitis, prolapse, calving difficulties.
Production: High beef yield, lean, flavorful, somewhat tough; average milk yield 12,000 lbs. annually, 3.7% protein, 4.2% butterfat.

The Simmental is an example of a breed whose dual-purpose capabilities are largely underutilized in America. It is used almost exclusively to produce crossbred beef calves in our country, but it still retains great dairy potential.

While this breed has many desirable characteristics, such as longevity, parasite resistance, adaptation to high altitudes, and a high tolerance of stress, it can be a challenge to maintain on forage alone.

It suffers from poor calving ease (although black Simmentals are somewhat better in this area due to Angus influence), tends to have frail calves, and requires a great deal of feeding to finish well.

White Park

Origin: Great Britain, ancient.
Breed Structure: Standardized.
Population: Endangered.
Uses: Beef, dairy.
Weight: Bulls 1,800–2,000 lbs., cows 1,200–1,300 lbs.
Preferred Climate: Adaptable to most climates.
Temperament: Very intelligent, wary; strong herd instinct, fiercely protective of calves; bulls can be dangerous.
Health: No major problems.
Production: Lean beef, exceptional taste, remarkably tender; moderate milk production.

The White Park once ran wild in the forests of Great Britain (in fact, feral herds still exist in the UK). It has changed little since that time, having retained its natural resilience and its uncanny survival instincts. While few breeds can equal the White Park in low-maintenance traits, its rather wild temperament is best left to those with ample experience with cattle.

For those willing to take on the challenge, however, the White

Park has much to commend it to the grass-based producer interested in raising gourmet beef. It is a hardy breed that can be counted on to take care of its own foraging, calving, and mothering duties with minimal assistance even in the most unfavorable environments.

CHAPTER 7
HONORABLE MENTION

The 35 breeds already covered are those most likely to find a place on your future homestead, farm, or ranch. However, there are a handful of less common breeds that people inquire about from time to time.

There are typically good reasons these cattle are not found on small farms. Some are difficult to raise due to health problems. Others are simply hard to come by. Either way, the honorable mention breeds are generally recommended for more experienced cattle owners.

But don't let that discourage you! If your research demonstrates that one of these breeds meets your five needs, go for it!

Belgian Blue

Origin: Belgium, mid-1800s.
Breed Structure: Standardized.
Population: Common.
Uses: Crossbred beef.
Weight: Bulls 2,400–2,750 lbs., cows 1,550–1,750 lbs.
Preferred Climate: Temperate.
Temperament: Extremely docile and affectionate.
Health: Birth defects, calving problems, underdeveloped reproductive tracts, laryngitis, bronchopneumonia.
Production: Extremely high yield, very lean meat, tender.

The Belgian Blue is known for double muscling, caused by a genetic mutation that prevents control of muscle development and leads to an increased number of muscle fibers. This mutation can create a host of health problems in purebred animals. Crossbred Belgian Blues carry only one copy of the double-muscling gene, which eliminates most of the health problems while still increasing the carcass yield.

Unfortunately, this large breed has extensive nutritional requirements and can rarely thrive on pasture alone. It requires a great deal of feed to finish well, but does produce a heavy, lean, high-value carcass with exceptional tenderness, high protein levels, and a healthy balance of essential fatty acids.

Breeding Belgian Blues is not a task for the faint of heart. This breed has a hard time mating without assistance and frequently requires C-section to give birth. However, it does make a dutiful mother.

Corriente

Origin: Mexico, late 1400s.
Breed Structure: Standardized.
Population: Secure.
Uses: Beef, rodeo.
Weight: Bulls up to 1,000 lbs., cows up to 800 lbs.
Preferred Climate: Hot.
Temperament: Gentle.
Health: No major problems.
Production: Small carcass, lean, flavorful.

The Corriente is a Spanish-derived breed that retains many of its historic low-maintenance traits, such as disease resistance, parasite resistance, foraging ability, willingness to consume weeds, fertility, and longevity. Its small frame may initially be seen as a disadvantage to the beef producer, but it does result in greater calving ease and less damage to pastures compared to many other beef breeds. Note that small size coupled with athleticism does translate into a bovine escape artist.

The Corriente is also the preferred breed for roping and many other rodeo sports due to its small size, its agility, and its fast-growing horns.

Piedmontese

Origin: Italy, mid-1800s.
Breed Structure: Standardized.
Population: Common.
Uses: Beef, crossbred beef, crossbred dairy.
Weight: Bulls average 1,850 lbs., cows average 1,250 lbs.
Preferred Climate: Adaptable to most climates.
Temperament: Calm, docile.
Health: Calving problems.
Production: Very high beef yield, exceptionally tender, lean.

Like the Belgian Blue, the Piedmontese carries the gene for double muscling. This results in large quantities of remarkably tender beef, but it does make it difficult for the cows to give birth. A common method of avoiding some of these problems while taking advantage of the carcass qualities of the Piedmontese is to breed Piedmontese bulls to cows of another breed.

Despite the drawbacks of double muscling, Piedmontese are hardy cattle overall, being quite tolerant of insects and extremes of temperature. They typically have long productive lifespans.

Pineywoods

Origin: United States, early 1500s.
Breed Structure: Landrace.
Population: Endangered.
Uses: Beef, dairy, draft.
Weight: Bulls 800–1,200 lbs.; cows 600–800 lbs.
Preferred Climate: Hot and humid.
Temperament: Varies by bloodline; ranges from active to docile.
Health: No major problems.
Production: Small carcass; average milk yield 5,000 lbs. annually, good butterfat.

This traditional American breed is highly adapted to the adverse conditions of the Deep South, being resistant to heat, humidity, parasites, and disease, and able to care for itself on poor-quality forage.

One remarkable trait of this landrace is its instinct to avoid predators by spending as little time as possible at a watering hole. The result is a breed with a smaller environmental footprint than is typical with most modern cattle—Pineywoods do not tend to pollute or destroy the banks of waterways.

A few Pineywoods cattle are dwarfs. Dwarfs are also known as "Guineas" in this breed.

Tarentaise

Origin: France, ancient.
Breed Structure: Standardized.
Population: Common.
Uses: Crossbred beef, dairy.
Weight: Bulls 1,600–2,100 lbs., cows 900–1,300 lbs.
Preferred Climate: Able to tolerate most harsh climates.
Temperament: Very docile.
Health: No major problems.
Production: High beef yield, lean, tender; average milk yield 12,000 lbs. annually, 3.4% protein, 3.6% butterfat.

Few Continental breeds are as well adapted to grass-based beef and dairy production as the Tarentaise. It is a hardy animal with low maintenance requirements and strong reproductive abilities. Coming from the French Alps, the Tarentaise is also well adapted to life at high altitudes, and it will frequently transmit this useful trait to its crossbred offspring.

The Tarentaise can consistently produce high-quality milk and meat on pasture alone. In France, it is exclusively used for making Beaufort cheese. In America, however, it is largely used for crossbreeding commercial beef calves.

APPENDIX
HELPFUL RESOURCES

Once you have determined what you expect from your cattle and have completed an initial survey of the breeds available, it is time for some more in-depth research. While a complete listing of breed organizations and their contact information would overwhelm this book and eventually become outdated, the resources listed under "Choosing a Breed" should still point you in the right direction. All of them offer additional information, as well as some combination of further reading material, breeder directories, and/or lists of breed associations.

There are also a few resources directed toward getting you off to a successful start with your first cattle. Many of these books and websites include additional breed profiles, as well.

Choosing a Breed
"Domestic Animal Diversity Information System (DAD-IS)"
Food and Agriculture Organization of the United Nations
http://www.fao.org/dad-is/browse-by-country-and-species/en/

This database offers profiles of numerous livestock breeds. Select "United States of America" as the country (assuming that is where you live) and "Cattle" as the species. Then click a breed that interests you to see more about its uses and characteristics.

"Cattle Breeds"
Homestead on the Range
https://homesteadontherange.com/cattle-breeds/

Our own online guide to cattle breeds including history, uses, temperament, health, and pros and cons.

"Cattle"
Breeds of Livestock
Oklahoma State University
http://afs.okstate.edu/breeds/cattle

This is a handy online encyclopedia-type reference packed with facts about both popular and rare cattle breeds. While little is known about some of the breeds, the compilers have made every effort to provide photographs and information on the history and characteristics of the cattle of the world.

An Introduction to Heritage Breeds: Saving and Raising Rare-Breed Livestock and Poultry
by The Livestock Conservancy
North Adams, MA: Storey Publishing, 2014

Exactly what the title suggests—an introduction, concise and clear enough for a reader with no prior experience with animals. Provides an overview of the process of maintaining heritage breeds in nontechnical terms. Also includes pithy breed profiles.

"Conservation Priority List"
The Livestock Conservancy
https://livestockconservancy.org/index.php/heritage/internal/conservation-priority-list#Cattle

This handy website covers only heritage cattle breeds, but it is a valuable repository of information. Click on a breed to find out more about its history, size, uses, temperament, and more. Helpful links are included to relevant breed associations and to a handy breed comparison chart.

Getting Your First Cattle

Grazing Systems Planning Guide
by Kevin Blanchet, Howard Moechnig, and Jodi DeJong-Hughes
St. Paul, MN: University of Minnesota Extension Service, 2000
https://conservancy.umn.edu/handle/11299/49821

This guide will help you evaluate current land and livestock resources before progressing to designing paddocks, fences, water systems, and heavy-use areas. Then comes a section on managing pastures for optimum performance, followed by a chapter on keeping records.

Waterers and Watering Systems: A Handbook for Livestock Owners and Landowners
by C.E. Blocksome and G.M. Powell (eds.)
Manhattan, KS: Kansas State University Agricultural Experiment Station and Cooperative Extension Service, 2006
https://www.bookstore.ksre.ksu.edu/Item.aspx?catId=361&pubId=1726

This free PDF download is packed with pros, cons, and design considerations for a number of water sources, power sources, and drink delivery options. Some additional material in the back of the book goes into the nuts and bolts of calculating well capacity, pipe diameter, and livestock water requirements, as well as obtaining permits when necessary.

Natural Cattle Care
by Pat Coleby
Austin, TX: Acres U.S.A., 2010

Key features of Coleby's nutrition-based program are pasture remineralization, a preventative stock lick, and vitamin- and mineral-based remedies for a host of ailments. Applies to both beef and dairy cattle.

The Jersey, Alderney, and Guernsey Cow
by Willis Pope Hazard
Philadelphia: Porter and Coates, 1872
https://books.google.com/books?id=I44OAAAAYAAJ

After examining the history of Channel Island breeds and how they were raised on their native islands, this book shifts to matters of importance for owners of dairy cows of all breeds. Interesting information is provided on choosing a good cow, feeding and housing her for her best health, and milking her properly.

How to Direct Market Your Beef
by Jan Holder
Beltsville, MD: Sustainable Agriculture Network, 2005
http://www.sare.org/Learning-Center/Books/How-to-Direct-Market-Your-Beef

In a friendly, easy-to-understand manner, Holder explains the different aspects of direct marketing that entrepreneurs will encounter, such as niches, pricing, distribution, advertising, public relations, and keeping records.

"Cattle"
Homestead on the Range
https://homesteadontherange.com/tag/cattle/

Our own growing collection of posts and resources on all aspects of cattle care. Topics cover both beef and dairy cattle and include everything from genetics to nutrition to draft work.

Salad Bar Beef
by Joel Salatin
Swoope, VA: Polyface, 1995

Explains the basics of raising grassfed beef, covering everything from buying your cattle to rotating pastures to direct marketing the beef. Upbeat and unconventional, this book challenges paradigms and inspires creativity.

Intensive Grazing: An Introductory Homestudy Course
by Burt Smith
Ridgeland, MS: Stockman Grass Farmer, n.d.

This power-packed bulletin condenses the most important points of management-intensive grazing into less than 20 pages. But don't think for a minute that the topic gets short shrift! This is a complete mini-course with information, a field exercise, a sample problem, and a quiz.

Storey's Guide to Raising Beef Cattle, 4th ed.
by Heather Smith Thomas
North Adams, MA: Storey Publishing, 2018

This guide presents all the basics of raising beef cattle, from choosing a breed to breeding a cow or heifer to weaning her calf. Includes information on fencing, housing, feeding, and health problems.

Veterinary Guide for Animal Owners, 2nd ed.
by C.E. Spaulding, DVM, and Jackie Clay
New York, NY: Skyhorse Publishing, 2015

Covers cattle, goats, sheep, horses, swine, poultry, rabbits, dogs, and cats. Each chapter begins with some information on basic care, then proceeds to a discussion of the health problems common to the species in question.

Reproduction & Animal Health
by Charles Walters & Gearld Fry
Austin, TX: Acres U.S.A., 2003

Learn how to "read" cattle to select sound, fertile breeding animals that will perform well on grass. Then find out how to keep cattle healthy by feeding the pastures. This is not a how-to guide for beginners, but a presentation of a philosophy, as well as an excellent starting point for further research.

MORE BOOKS FROM HOMESTEAD ON THE RANGE

The Family Garden Journal: A Keepsake of Daily Plans, Observations, and Harvests
by Michelle Lindsey

The Family Garden Journal offers an effective way to make the most of your learning experience in the garden, empowering you to become a true green thumb. Hundreds of carefully designed and tested pages will give you room to plan and observe your garden, every day of the year, including a step-by-step gardening guide, a shopping list, a garden map, a planting schedule, a family workload table, attractive reference pages, and a 366-day journal.

2nd edition. 466 pages. ISBN-13 (paperback) 978-0997526103.

The Worst Jokes I Know (and I Know a Lot!): 101 Funny Bone Ticklers for Jokesters of All Ages
by B. Patrick Lincoln

At last—a supply of clean jokes for the whole family! Young jokers (cards, you might say) will learn 101 side-splitting riddles and funny bone ticklers, all in a small, illustrated book made to be shared. Jokesters of all ages will enjoy trying out corny wordplay on their family, friends, and pun pals. These clean jokes and riddles are suitable for children ages 4 through 104. After all, you're never too old for a good joke. Share a laugh!

62 pages. ISBN-13 (paperback): 978-0997526110. ISBN-13 (eBook): 978-0997526127.

Available at Amazon.com. Find a complete list of titles at HomesteadOnTheRange.com/our-books/.

Made in the USA
Coppell, TX
02 January 2020